CREATION COLLECTION

ANIMALS
How Creature Features Defy Evolution

CREATION COLLECTION

ANIMALS
How Creature Features Defy Evolution

Frank Sherwin and Jeffrey P. Tomkins, with Brian Thomas, Randy J. Guliuzza, and James J. S. Johnson

Dallas, Texas
ICR.org

Animals
How Creature Features Defy Evolution

Contributions by Frank Sherwin, D.Sc. (Hon.), and Jeffrey P. Tomkins, Ph.D., with Brian Thomas, Ph.D., Randy J. Guliuzza, P.E., M.D., and James J. S. Johnson, J.D., Th.D.

First Printing: August 2024

Copyright © 2024 by the Institute for Creation Research. All rights reserved. No portion of this book may be used in any form without written permission of the publisher, with the exception of brief excerpts in articles and reviews. For more information, write to Institute for Creation Research, P. O. Box 59029, Dallas, TX 75229.

Designed by Melissa Marquez

Concept by the ICR Communications team: Jayme Durant, Beth Mull, Susan Windsor, Michael Stamp, Lori Fausak, Renée Dusseau, Bethany Trimble, Rachel Brown, Frank Sherwin, and Sydney Walters

All Scripture quotations are from the New King James Version.

ISBN (paper): 978-1-957850-68-9
ISBN (ebook): 978-1-957850-74-0
Library of Congress Control Number: 2024942559

Please visit our website for other books and resources: ICR.org

Printed in the United States of America.

Published by Wayfinders Press, an imprint of ICR Publishing Group

Cover: Hummingbird (*Archilochus colubris*)

TABLE OF CONTENTS

Introduction 7

Complexity Calls for a Creator
   1. Architecture and Engineering in Created Creatures 9
   2. Complex Creature Engineering Requires a Creator 18
   3. Intricate Animal Designs Demand a Creator 23
   4. Embryonic "Clocks" Mimic Human Construction Schedules ... 29

Flying Creatures
   5. Honeybee Design Saves Energy 35
   6. The Passive Stealth Wing of the Moth 38
   7. Fruit Fly Jitters 42
   8. Aerial Engineering and Physics of the Dragonfly 47
   9. Open Ocean Dragonfly Migration Boggles the Mind 50
  10. Why Don't Raindrops Bomb Butterfly Wings? 53
  11. Butterfly Wing Design Repudiates Evolution 56
  12. Hummingbirds by Design 60
  13. Hummingbird Flight Strategies 73
  14. The Syrinx Song 77

Aquatic Animals
  15. Created Cuttlebone's Sweet Spot 82
  16. Molecular Motors of a Squid Show CET in Action 86
  17. Marine Sponges Inspire 91
  18. Clever Construction in Rorqual Whales 95
  19. Brittle Stars See with Their Skin 99
  20. Common Seals Display Extraordinary Bioengineering 103

21. Complex Metabolic Process in Fish Startles Evolutionists 109
22. Fish Body Design Reveals Optimized Swimming Mechanics .. 112
23. Does Oddball Platypus Genome Reveal Its Origins? 115

Life on Land

24. Beetle Mouth Gears Shout Design 119
25. Termite Nest Architectural Design Is Clearly Seen 123
26. Improved Steel Copies Bone Microstructure 127
27. Design-Based Spider Research Proves Creator's Genius 130
28. Spiders Have Built-In Algorithm to Construct Webs 135
29. Do Shrinking Shrews Cheat Evolution? 138
30. Hero Shrew Spine Design Glorifies the Creator 142
31. Geckos Have Holes in Their Heads 145

Contributors 149
Image Credits 151

INTRODUCTION

Our beautiful Earth is filled with millions of animal species, and countless more remain to be discovered. According to conventional science, living things are the result of evolution through random mutations, lucky products of time and chance. Evolutionary theory asserts that the species we see today evolved from a common ancestor and diversified over millions of years. But does this actually explain the incredible variety and apparent design of life?

In *Animals: How Creature Features Defy Evolution*, ICR scientists and experts test this theory and find that the evidence demonstrates that only God could have crafted the wondrous beauty and complexity of animal features and functions. Each creature is engineered to thrive in its habitat and is equipped with traits that work together precisely for its survival. The study of animals' biological abilities has inspired centuries of manmade inventions, and it's clear they were specially and purposefully designed.

Living creatures not only defy conventional thinking but also testify to the greatness of their Creator. God's creation proclaims the majesty of His handiwork. "O Lord, our Lord, how excellent is Your name in all the earth!" (Psalm 8:9).

1
ARCHITECTURE AND ENGINEERING IN CREATED CREATURES

Frank Sherwin, D.Sc. (Hon.)

"Nature is a pretty impressive engineer," stated evolutionist Daniel Lieberman in *Nature* magazine. He noted:

> The physical world poses many basic challenges, such as gravity, viscosity and pressure gradients, to all living creatures, which in turn have evolved an astonishing array of solutions. Many of these, such as paddles, valves and hydrostats, are so widespread that we rarely notice them. Others perform so well that we marvel at their superiority to human-made devices.[1]

Creationists maintain it was God who addressed these basic challenges with astonishing solutions—not chance evolutionary processes working for millions of years. Indeed, even if we were to give more time than what the evolutionists estimate, we would still never see nature producing animals and their multiple systems with such superior function and detail.

Biomechanics is the field of biology that studies the action of internal and external forces on the living body, especially the skeletal system.[2] Also called bioengineering, this fascinating area analyzes biological design and the physical forces associated with humans and animals. If ever there was evidence for creation on a macroscopic scale (Romans 1:20), it would be the vast array of creatures all over the world that are marvelously designed to move in and fill their environments based on these amazing design features.

We see that when architects and engineers design buildings or other structures, they either knowingly or unknowingly follow the Creator's design features found in human anatomy and the animal world. Most of us are aware of the beautiful medieval cathedrals of Europe. Built to the glory of God, they are also a testament to plan, purpose, and detailed design that leaves nothing to chance. The success of such efforts is seen in their victory over gravity through the centuries.

The architectural planning and detail of a typical cathedral are stunning. A majority of large churches and cathedrals in Europe are designed with a wide nave, the central aisle of a basilica church. They also tend to have a lower

Figure 1. Salisbury Cathedral

Figure 2. Cloister in Gloucester Cathedral

aisle separated by an arcade (a succession of arches) on either side. The most prominent external feature of the building is the spire, built for vertical emphasis. The tallest spire in England (404 feet) is found on Salisbury Cathedral, mostly built during the years 1220 to 1258 in the Early English Gothic style (Figure 1).

Medieval architects followed design patterns in much the same way as the divine Architect who exquisitely designed the bones and skeletons of humans and animals. We see that bone construction is designed for the needs of the person or animal in question. Bone is strongest when burdened in compression but weakest in sheer and tension. Any architect would look at the walls and ceiling of, for example, Gloucester Cathedral (c. 1355, Figure 2) in England and think of pneumatic bird bones if they'd had the opportunity to view that internal avian structure.

Figure 3. Bird bone lattice structure

Bird bones are not solid like those one would find in dinosaurs, such as 30-plus-ton sauropods. Compressive and tensile stress put on the bird's skeleton during flight is largely handled by the solid surface of the bone and less in the central portion. Not surprisingly, the interior of the bone is designed with a web of supportive struts (Figure 3) that look much like what the Warren truss uses (Figure 4).

The basic structural elements of a Gothic cathe-

Figure 4. Warren truss bridge

dral are the interior and exterior piers (the sidewalls seen from the outside) with critical flying buttresses in between (Figure 5). Medieval architects designed flying buttresses to resist bending of the main interior piers due to lateral pressure caused by snow or wind. The buttresses support the structure by carrying lateral thrust from the upper areas of the building all the

Complexity Calls for a Creator

Figure 5. Flying buttresses on St. Vitus Cathedral

way to the ground.

We see from a person's skeletal design that the heads of the two femurs carry the weight of the upper body (Figure 6). During a striding gait, the head and neck of one femur carries all the weight of the upper body. This explains why the elderly who suffer from osteoporosis often break one of their hips—fracturing the neck of the femur—and *then* fall down. It rarely happens the other way around. The shaft of the femur, which is the longest bone in the body, experiences an asymmetric load that increases the propensity to bend. But there is a long tendon on the lateral side of the thigh called the iliotibial tract, and it is so strong that it serves as a site of insertion for the tensor fasciae latae muscle. Such design helps to counteract the bending forces on the femur. The tensile stress (such as that addressed in cathedral construction) that develops on the lateral side of the femur is counteracted.

Loading is a problem also dealt with in the design of both man-made structures and people. God de-

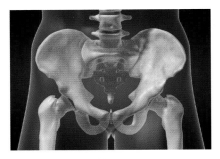

Figure 6. Skeletal hips

signed people and most animals with bilateral symmetry—an arrangement such that one plane divides the person in two halves that are approximate mirror images. Symmetric loading of supportive columns, such as the human spinal column, results in a centered compressive force in an interior pier of a cathedral or an upright human.

Since the weight of the upper body is above the hips, there is potential instability in upright posture. God designed features of our skeletal anatomy that contribute to stability. Our spinal column is designed with vertebrae stacked above the hips, much like the main mast of a cutter or schooner. The complex system of ropes, called rigging, that supports a ship's mast and sails is much like the muscles and ligaments specifically aligned to the lower vertebrae and ribs that support the spinal column. One is struck by the similarity of a cutter's rigging (looking from the stern to the bow) and the ligaments and muscles supporting the vertebrae. In addition, God has designed human pelvic anatomy to have a broad base of support of the upper body by the shortening and flaring of the pelvis.

Four-legged animals (tetrapods) must have a specialized spinal column that helps suspend the weight of the body. Again, the field of engineering allows us to appreciate just what is happening as weight is distributed between the front and back legs. A typical suspension bridge is composed of tension members and compression members. Bridge builders design these structures so the rigid members resist compression and the more flexible members resist tension.

Sections or spans of the bridge are designed to rest on piers (supports for the ends of adjacent spans). Combining the piers and spans allows the weight of the bridge to span the distance between the piers. Not surprisingly, the area midway between the two piers (the nodal) is the weight distribution trade-off.

A tetrapod vertebral column has approximately the same function. The nodal point depends on the weight distribution of the front and back legs. Neural spines of the vertebrae reverse their orientation at the nodal of the spinal column. The centra and spines of a vertebra are like the compression members, while the muscles and ligaments are like the tension members. The ligaments and muscles of the creature are designed to principally resist tensile forces, while the bones resist compression.

Evolutionist Michael Benton discusses the functional morphology of the pliosaur *Rhomaleosaurus*:

> In pliosaurs, the jaw was designed to clamp shut with huge force, and to prevent the prey struggling free. The shape of the pliosaur jaw, with an elevated coronoid eminence near one end has been compared to an asymmetrical swing bridge that is loaded by its own weight when it is open.[3]

Solutions to mechanical stress in animals are also seen in the design of the arch, a structure that is curved or bowed as seen, for example, in an arched suspension bridge. Arches work in engineering as long as they maintain their shape and don't flatten out. The weight of the bridge is maintained as long as the arch above it is upheld. Looking at God's mammals, we see the sternum, ligaments, and abdominal muscles are designed to maintain the arched nature of the vertebral column, much like the roadbed between the piers of an arched suspension bridge.

Trunk vertebrae of tetrapods resemble an archer's bow. Such bow design is also seen in the neck of sauropods, where the brachiosaur's cervical ribs can be compared to leaf springs.[4]

No one would suggest such magnificent constructions as cathedrals, schooners, or bridges are the result of random natural processes. How much less the human body? When viewing and addressing God's design of people and animal creatures, the term engineered morphology comes to mind—looking at human and animal anatomy from a biomechanical viewpoint. But there is also ecological morphology, viewing an animal in its natural environment as it purposefully migrates, not randomly mutates. Animals are designed to move in and fill an environment—which is ecology, not evolution. God engineered animals with the innate ability to adapt, which sometimes even leads to speciation. Finally, creation morphology brings together this information for us to observe, measure, and research God's creatures using the perspectives of function, form, ecology, and design. The more we learn, the greater our awe in the Master Designer who made it all.

References

1. Lieberman, D. E. 2004. Engineering for animals. *Nature.* 428 (6986): 893.
2. Sherwin, F. 2002. God's Creation Is "Clearly Seen" in Biomechanics. *Acts & Facts.* 31 (3): 4–5.

3. Benton, M. 2015. *Vertebrate Paleontology,* 4th ed. Malden, MA: Wiley Blackwell, 34.
4. Thomas, B. Amazing Sauropod Neck Design in "Cervical Ribs." *Creation Science Update.* Posted on ICR.org November 5, 2015, accessed August 15, 2017.

2
COMPLEX CREATURE ENGINEERING REQUIRES A CREATOR

Jeffrey P. Tomkins, Ph.D.

High-Tech Bat Sonar

Bats are a remarkable example of God's handiwork. Their sonar capabilities especially put anything human-engineered to shame.[1,2] These creatures appear suddenly in the fossil record in Eocene strata with no evolutionary precursors, and their fossils look just like modern bats. Bats are the only mammals capable of true, sustained flight like birds. In fact, they are even more maneuverable in the air than most birds are.

Bats use an incredibly complex form of echolocation to locate prey in the dark. As they zip through the air, they constantly emit and sense sound waves to accurately pinpoint the exact locations of moving targets, which they then snatch out of the air and eat completely "on the fly." The variable sound pulses bats send out have been measured at 30,000 to 100,000 hertz (Hz).[3] In comparison, the upper bound for human hearing is 20,000 Hz.

This high-tech sonar is amazing enough, but bats

Pipistrelle bat

also contain another very interesting piece of engineering. They would actually deafen themselves if it weren't for a highly specialized inner ear muscle. This muscle contracts rapidly, repeatedly, and precisely to "freeze" the bone associated with hearing exactly when the sonar impulse is sent out. Then it relaxes at just the right time to receive the incoming sonar echo information from previous impulses.

This echolocational system is so precise that bats can use built-in neurological algorithms to intuitively process the ultrasonic sonar pulses to "see" their surroundings with sound just as well as people can see with their eyes! Some bats can even target and nab insects as small as a mosquito.

Honeybee Waggle Dancing

One aspect of honeybees that fascinates scientists is their eusocial (cooperative interaction) behavior, especially when it comes to locating food and other resources and then communicating that highly spe-

Honeybees

cific information to their hive mates on their return.[4] When a foraging bee discovers a new food or water source, it flies back to the hive and conveys the exact coordinates of the resource through a high-tech waggle dance in a figure-eight pattern. The angle of the dance in relation to the sun confers the direction, while the amount of waggling confers distance and the general utility of the resource (e.g., food or water).

The hive mates surrounding the dancer bee also exhibit highly specific behavior that involves their distance and angle in relation to the dancer. An important part of their engagement in the information acquisition process is touching antennae with the dancer bee. Located in the bee's antennae are highly specific, ultrasensitive mechanosensors that detect the information-rich vibrations from the dancer bee in a range of 265 to 350 Hz. Research has shown that the bees emit several different detectable chemical signals during the waggle dance, as well. Sophisticated behavioral communication like this in a seemingly "simple" insect utterly defies evolutionary myth and glorifies our omnipotent Creator.

Monarch Butterfly Navigation

Monarch butterflies' annual long-distance migrations are yet another example of the Creator's genius. These insects accurately navigate a southwesterly course on a 2,400-mile autumn trip from Canada and the northern U.S. to specific sites in Mexican forests.[5] Part of this extraordinary journey can take the butterflies across hundreds of miles of open ocean in the Gulf of Mexico. The butterflies navigate the whole journey by continuously tracking data with their eyes on the horizontal position of the sun over the course of the day.

Researchers have also discovered the butterflies have a time-compensation clock located in their antennae to aid in decoding the sun's movements in relation to time—also known as a circadian clock. In any man-made system, this would require complex sensors, computer algorithms, and hardware to de-

The majestic monarch butterfly

code and integrate the data as part of the overall navigation and flight system. God's design in this small insect and other creatures puts man's efforts to shame.

References

1. Sherwin, F. 2003. Bat-tastic Bats. *Acts & Facts*. 32 (10).
2. Sherwin, F. 2015. The Ultrasonic War Between Bats and Moths. *Acts & Facts*. 44 (10): 15.
3. A hertz equals one vibration cycle per second.
4. Collison, C. A Closer Look – Waggle Dances. *Bee Culture*. Posted on bee-culture.com April 23, 2018.
5. Guliuzza, R. J. 2018. Engineered Adaptability: Creatures' Adaptability Begins with Their Sensors. *Acts & Facts*. 47 (3): 17–19.

3
INTRICATE ANIMAL DESIGNS DEMAND A CREATOR

Jeffrey P. Tomkins, Ph.D.

Evidence of our Creator is all around us. Romans 1:19–20 states, "What may be known of God is manifest in [people], for God has shown it to them. For since the creation of the world His invisible attributes are clearly seen, being understood by the things that are made." God's handiwork is certainly manifested in the exquisite, engineered design of His creatures.

The Salmon

One example of God's creative genius is the salmon. This fish is born inland in freshwater streams that are miles from the ocean, migrates to live in the salty sea, and then returns to fresh water so it can spawn. The salmon has a unique ability to maintain a constant healthy level of saltiness. Its internal cellular and organellar systems adjust automatically in response to environmental tracking systems that monitor external salt levels.[1]

Chief among these engineered systems are specialized sodium pumps embedded in the cell membranes. The pumps' activity is coordinated within the internal apparatus of the cell and also with other systems in the salmon's various organs, especially those on the forefront of osmoregulation (the maintenance of body-fluid pressure) such as the gills and kidneys.

Salmon

In addition to these integrated cellular systems, the salmon has built-in behavioral traits to manage its salt levels. Instead of immediately charging into the ocean or back into fresh water, it pauses to temporarily equilibrate its body in transitionary zones between the two.

The Dragonfly

The field of bioengineering makes use of the design found in living creatures. One flying creature human engineers have tried to copy is the dragonfly. These insects are expert fliers. They can maneuver straight up and down, hover in place like a helicopter, and even mate in midair.

The dragonfly's optics are also amazing with almost its entire head composed of visual sensors loaded with engineering that's only beginning to be understood. It has very complex eyes constructed of individual visual sensory units called ommatidia. A single compound eye has an integrated lens system containing up to 30,000 ommatidia. Each individual ommatidium collects its own stream of visual information that's transmitted to the dragonfly's brain,

Dragonfly

where it's decoded and processed to form a mosaic image with intricate visual depth and detail.

Combined with its flight capabilities, the dragonfly's high-tech visual system allows it to track and grab aerial targets like flies with deadly precision. A study of caged dragonflies found they were able to successfully snatch their rapidly moving prey out of the air with 95% accuracy.[2]

The Hummingbird

The hummingbird is another animal that glorifies the Creator. This little creature is distinctly different from all other bird kinds. Hummingbirds are the only birds that can fly backward. They can literally zip around in just about every direction due to their wings' ability to rotate in a full circle and flap up to 80 times per second.

Hummingbirds have much larger and more complicated brains than insects. One study determined that "the hummingbirds had faster reaction times than those reported for visual feedback control in insects."[3] The endurance and speed of

Animals: How Creature Features Defy Evolution

Hovering hummingbirds

hummingbirds are also phenomenal. They can fly at 25 to 30 mph and dive at speeds of up to 60 mph. The ruby-throated hummingbird can travel up to 500 miles across the Gulf of Mexico to reach its breeding grounds.

The V-22 Osprey is human engineers' attempt to create something with roughly similar flight capabilities. This military aircraft requires constant fueling, maintenance, supervision, and construction since it can't feed itself or reproduce like the hummingbird. Human efforts to copy the intricate form and function of God's creatures is further evidence of our Creator's engineering genius.

References

1. Vargas-Chacoff, L. et al. 2018. Effects of elevated temperature on osmoregulation and stress responses in Atlantic salmon *Salmo salar* smolts in fresh water and seawater. *Journal of Fish Biology.* 93 (3): 550–559.
2. Gonzalez-Bellido, P. T. et al. 2013. Eight pairs of descending visual neurons in the dragonfly give wing motor centers accurate population vector of prey direction. *Proceedings of the National Academy of Sciences.* 110 (2): 696–701.
3. Cheng, B. et al. 2016. Flight mechanics and control of escape manoeuvres in hummingbirds. I. Flight kinematics. *Journal of Experimental Biology.* 219 (22): 3518–3531.

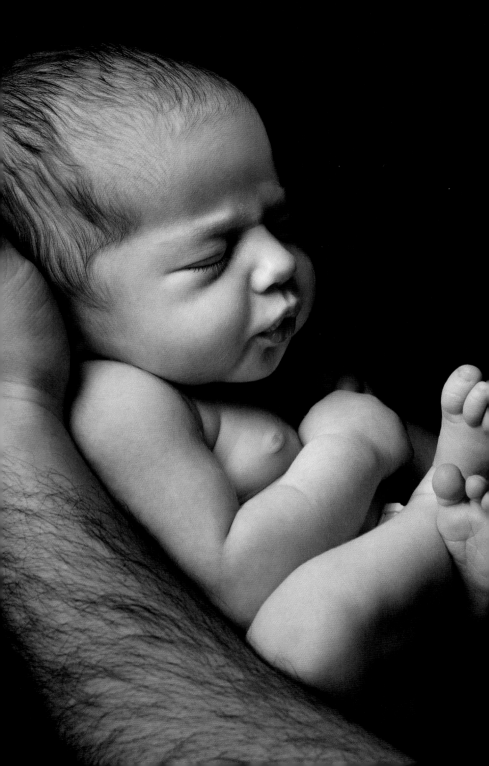

4
EMBRYONIC "CLOCKS" MIMIC HUMAN CONSTRUCTION SCHEDULES

Randy J. Guliuzza, P.E., M.D.

Two recent findings in biology add confirmation that biological functions are best characterized by engineering principles. This research describes a number of sophisticated internal clocks that control the timing of key events during embryological development. These clocks are part of systems that function just like a construction schedule used to guide decisions by human project managers.

ICR has been developing a theory of biological design (TOBD) that is a better scientific explanation for biological functions than selectionism.[1] The theory is a working hypothesis of how the biological phenomena of growth, metabolism, reproduction, and adaptation function. One major premise is that biological functions are suitably explained by engineering principles. We expect to uncover elements in biological systems that correspond to human-designed systems performing similar functions. Thus far, our theory has guided the development of several organism-focused, design-based models. One model, called continuous environmental tracking (CET), explains biological adaptation;[2] another explains seed dormancy and germination in forest ecosystems;[3] and a third model explains that the human immune system does more than just defend the body against disease—it has a much broader purpose of acting as an interface that regulates the relationship primarily between organisms and the microbial realm.[4]

Both embryological development and human-engineered building projects assemble materials into a final pre-determined product. It would seem, therefore, that a design-based framework to model embryological development could be readily created. It is true that intra-cellular information and DNA do correspond to the plans and specifications produced by human engineers in many ways.

But the construction schedule itself is another product generated by engineers that is of equal importance to the plans for completing a project. A theory of biological design expects that the biological equivalent of a construction schedule for embryological development exists in the genetic information of the first single cell. Fortunately, research is beginning to identify in creatures the parts of logic-based mechanisms that function as schedules.

Olivier Pourquié, professor of genetics in the Blavatnik Institute at Harvard Medical School, has built on his pioneering discovery of a cellular "clock" in chicken embryos.[5] One area of his research focuses on how development in vertebrates is controlled. The spine is both a benchmark and foundational fea-

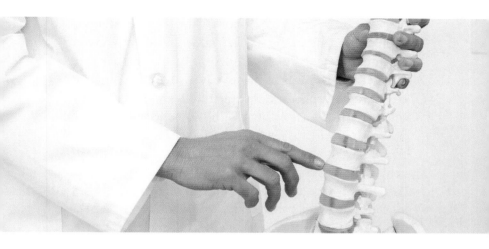

ture built very early in vertebrate embryos. Per the Harvard press release, Pourquié's "clock" does in fact schedule the timing of spine development "where each "tick" stimulates the formation of a structure called a somite that ultimately becomes a vertebra."[5] Pourquié confirmed that human development is also controlled by the same type of clock and said, "Our system should be a powerful one to study the underlying regulation of the segmentation clock."[5]

Research led by James Briscoe, a developmental biologist at the Francis Crick Institute in London, compiled information on multiple scheduling mechanisms. These were recounted in a special report by *Knowable Magazine*.[6] Stressing the importance of these mechanisms within embryos, the report begins by saying, "If events weren't properly timed, pandemonium would ensue. Researchers are ferreting out the internal clocks that control developmental sequence and scheduling."

Briscoe intended to discover "the timers that keep things trucking along at the right rate for any given organism, ensuring that it grows to the proper size and with all its parts in place." Thus, an embryological construction schedule is expected to control the timing and rate of assembly just like a schedule used by human engineers. *Knowable Magazine* summed up Briscoe's explanation: "First of all, for bodies to properly form, events must unfold in the right sequence: A before B, and B before C, and so on, at right times all over the developing body," while a "second aspect of timing is far more mysterious: the molecular processes setting the tempo such that clocks run faster or slower in different species."[6]

Similar to Pourquié's research, Briscoe identified different types of clocks in different tissues, which helped provide details on how they operated. They found elaborate, innate systems that facilitate very different types of operation. Some clocks may "count up," "count down," run through feedback loops resulting in "cyclical oscillations," or keep track of how many times they've divided.[6] *Knowable Magazine* quotes developmental biologist Mubarak Hussain Syed of the University of New Mexico as saying, "The cell might be counting, 'OK, we have done 20 divisions, and now it's time' for the next step."[6]

What makes everything even more remarkable is that, unlike current human-engineered devices that are assembled by humans or robots, organisms self-assemble. A human construction manager is constantly making logical decisions to keep the project efficiently moving toward completion based on the specifications, schedule, and many changing conditions impacting the project. But suppose that engineers wanted to build a device that could actually self-assemble. Obviously, the engineers must program into this device the specifications for all of its traits. But they must also program an incredibly complicated set of logic-based responses that will imitate their conscious logical intentions as if they were on site to either direct assemble per the schedule or select between pre-programmed potential solutions to different challenges. The information content to implement *control*—including the timing of the assembly—could greatly exceed the information content for the *traits*. To substitute for the project engineers' eyes and ears, a self-assembling device would also need a large array of internal and external sensors to send data to the internal, logic-based controller.

NASA's R5 robot "Valkyrie"

As the schedules controlling embryonic development become clearer, we are closer to modeling development with the same engineering principles that control human construction projects. This model will be able to make predictions of future discoveries in developmental biology and could guide research. A fresh new benefit to our theory of biological design is its ability to generate these organism-focused, engineering-based models.

Innate control systems are essential to all dynamic entities, and they are a hallmark of sophisticated engineering. The only known originating source of this information is a real conscious mind. The myriads of phenomenally complex control and backup systems are evidence of the astounding creative genius of nature's Creator, the Lord Jesus Christ.

References

1. Guliuzza, R. J. 2018. Engineered Adaptability: Adaptive Changes Are Purposeful, Not Random. *Acts & Facts.* 47 (6): 17–19.
2. Guliuzza, R. J. and P. B. Gaskill. 2018. Continuous environmental tracking: An engineering framework to understand adaptation and diversification. In *Proceedings of the Eighth International Conference on Creationism.* J. H. Whitmore, ed. Pittsburgh, PA: Creation Science Fellowship, 158–184.
3. Hennigan, T. and R. Guliuzza. 2019. The Continuous Environmental Tracking hypothesis—application in seed dormancy and germination in forest ecosystems. *Journal of Creation.* 33 (2): 77–83.
4. Guliuzza, R. J. and F. Sherwin. 2016. Design Analysis Suggests that our "Immune" System Is Better Understood as a Microbe Interface System. *Creation Research Society Quarterly.* 53 (5): 123–139.
5. Dutchen, S. 2020. Backbone of Success. Harvard Medical School News & Research. Posted on Harvard.edu January 8, 2020, accessed on February 4, 2020.
6. Brown, E. 2020. How does the embryo make all its parts at just the right moments? *Knowable Magazine.* Posted on knowablemagazine.org on January 30, 2020, accessed on February 4, 2020.

5
HONEYBEE DESIGN SAVES ENERGY
Frank Sherwin, D.Sc. (Hon.)

Biomimicry is the making of systems or materials that are modeled after flora or fauna found in God's creation (e.g., the artificial fabric Velcro is modeled after burrs).

Scientists have uncovered and learned from many creatures in God's creation—for example, from a host of insects[1] and especially the bee and its honeycomb design that has inspired the production of insulation, aircraft parts, and cardboard boxes.

It is not surprising that entomologists have yet again learned from this industrious insect, this time in regard to minute hairs on the honeybee that reduce friction.[2] This design clearly shows[3] God's engineering in living systems.[4]

When observing the honeybee, one can see the constant straightening and curving of its abdomen—this creates friction of the outer plates of the exoskeleton. One would expect that such movement over time would cause significant wear. However, there is very little, and a group of scientists wanted to know why. Using an electron microscope, they viewed

the abdominal plates and were surprised to see tiny-branched hairs. They hypothesized these may have something to do with friction reduction and set up a series of experiments measuring abdominal segment movement with and without the hairs.

> As the load increased, friction for the hairless surface rose, whereas no obvious rise in friction was observed for the hairy surface. The researchers calculated that the hairy surface reduced abrasion during abdominal contraction by about 60% and also saved energy with each contraction. This adds up to a large amount of conserved energy that is essential for conducting bees' daily activities, the researchers say.[2]

This discovery could be incorporated by bioengineers in designing longer-lasting moving parts that "someday [could] be used to extend the lifetime of engineered soft devices, such as actuators and hinges."[2]

One cannot help but be reminded of the words in Proverbs 6, "Look to the ant, thou sluggard! Consider her ways and be wise."[5] Of course, not all scientists are sluggards, but they would do well to continue to investigate and possibly mimic the intelligent design found in the insects in God's world.

References

1. Sherwin, F. 2007. The Amazing Jewel Beetle. *Acts & Facts.* 36 (5).
2. American Chemical Society. Honeybees' hairy abdomens show how to save energy, reduce wear on materials. *Phys.org.* Posted on phys.org June 9, 2021.
3. Romans 1:20.
4. Tomkins, J. P. 2015. Optimized Design Models Explain Biological Systems. *Acts & Facts.* 44 (2): 14.
5. Proverbs 6:6.

6
THE PASSIVE STEALTH WING OF THE MOTH

Frank Sherwin, D.Sc. (Hon.)

We appreciate the beautiful butterfly but not so much the pesky moth. However, the wing structure of both creatures is amazing, "The wings of moths and butterflies are densely covered in scales that exhibit intricate shapes and sculptured nanostructures."[1]

In addition, much research has been conducted regarding the remarkable abilities of the bat to navigate and hunt in the dark for tasty moths.

> Bats use an incredibly complex form of echolocation to locate prey in the dark. As they zip through the air, they constantly emit and sense sound waves to accurately pinpoint the exact locations of moving targets [such as the moth], which they then snatch out of the air and eat completely "on the fly." The variable sound pulses bats send out have been measured at 30,000 to 100,000 hertz (Hz). In comparison, the upper bound for human hearing is 20,000 Hz.[2]

But as effective as the bat sonar is, it has been found that some species of nocturnal moths (e.g.,

Chinese silk moth

the Chinese silk moth, *Antheraea pernyi*) are able to evade its highly sophisticated tracking system due to tiny scales on their wings.

Scientists have discovered the surface of moth wings have been designed with an incredible sound-absorbing surface called metamaterial.[3] They believe that this is the first acoustic metamaterial found in nature and that it actually creates "acoustic camouflage against bat echolocation."[1] The wing surface functions "as efficient sound absorbers through the action of the numerous resonant scales that decorate their wing membrane."[3]

One can only be astounded by the high degree of extremely small (or nano-) engineering that went into covering the wings with different scale sizes that have different resonant frequencies, giving broadband acoustic absorption (i.e., the scales take in and "hold" the sonar pulses of the bat, as opposed to the pulses being reflected back).

The wings of moths are decorated with scales of varying size, each with its own resonant frequency. Each scale absorbs sound at the frequency of its main resonance modes. When numerous scales of differing size and therefore resonant frequencies cover the membrane, the result is broadband acoustic absorption in the deep-subwavelength regime.[3]

What an altogether unique way to ensure virtually all phases of bat sonar would be absorbed (about three octaves of sound), therefore making the moth virtually invisible to the bat!

How did such amazing acoustic camouflage come about in these moths? Non-evolutionists should not be intimidated when phrases such as "have evolved" are used. Shen et al. stated, "The scales on moth wings have evolved to reduce the echo returning to bats."[1] But it is just as scientific to say, "The scales on moth wings are *designed* to reduce the echo returning to bats." Furthermore, statements such as, "In all likelihood, this natural metamaterial has been sculpted by millions of years of evolution"[4] can be adjusted to say, "In all

Luna moth wing scales

likelihood, this natural metamaterial has been sculpted by the hand of the Creator."

Not surprisingly, "there has been a growing interest in bioinspired metamaterials, with researchers looking to nature for clues into designing the next generation of advanced metamaterials."[3] Scientists appreciate such microengineering, but they are attempting to copy what God—not nature—has designed in the moth just thousands of years ago.

References

1. Shen, Z. et al. 2018. Biomechanics of a moth scale at ultrasonic frequencies. *Biophysics and Computational Biology*. 115 (48): 12200–12205.
2. Tomkins, J. Complex Creatures Engineering Requires a Creator. *Creation Science Update*. Posted on ICR.org July 31, 2019, accessed August 17, 2022.
3. Neil, T. et al. 2022. Moth wings as sound absorber metasurface. *Proceedings of the Royal Society A*. 478 (2262).
4. Cassella, C. Moth Wings Have Evolved a Rare 'Metastructure' We've Been Trying to Make in The Lab. *ScienceAlert*. Posted on sciencealert.com November 28, 2020, accessed August 17, 2022.

7
FRUIT FLY JITTERS
Frank Sherwin, D.Sc. (Hon.)

Researchers working with fruit flies—the ubiquitous lab animal—have discovered the flies are able to undergo an amazing ocular process called microsaccades (involuntary microscopic jittering of their eyes).

This means that as one constantly stares at an object, it won't fade in the same way the olfactory nerves in our nose may be overcome or "go blind" after time with a constant odor. This is because your eyes undergo these subtle movements, giving "just enough variety in the light patterns on your eyes to prevent your visual neurons from completely adapting to what they're looking at."[1]

But, as stated in the Rockefeller University's news article,

> Insects don't have this luxury because their eyes are fixed firmly to their heads. But a new study shows that fruit flies have evolved a different strategy to adjust their vision without moving their heads—they move the retinas inside of their eyes.[1]

Creationists would say the fruit fly was *created* with a different strategy to adjust their vision.

Flies and humans were created separately and therefore have very different types of eyes, but it's hardly surprising that they have "similar active strategies" such as the ability to track.[1] One scientist was quoted as saying, "We think the ability to track moving objects evolved independently in flies and humans,"[1] but this isn't evolution via an unknown common ancestor of insects and people.[2] This is common design by the Creator, who used the same engineering principles to achieve the same function for different creatures.

The research the scientists conducted is top notch, catching key details:

> Here we show that *Drosophila* use their retinal muscles to smoothly track visual motion, which helps to stabilize the retinal image, and also to perform small saccades when viewing a stationary scene. We show that when the

Fruit fly eye

retina moves, visual receptive fields shift accordingly, and that even the smallest retinal saccades activate visual neurons.³

However, the evolutionists fall back on the unsatisfying evolutionary convergence argument. "The similarities of smooth and saccadic movements of the *Drosophila* retina and the vertebrate eye highlight a notable example of convergent evolution."³ Convergent evolution is a theory that tries to address resemblance in distinct evolutionary lineages. The theory is used to explain why, for example, the European Timber wolf and Tasmanian marsupial wolf look virtually identical even though they're on widely separated continents. As the late Tom Bethell said, convergence "undermines the assumption that similarity is the result of common ancestry."⁴

The Creator's hand is seen in the fruit fly eye design that superficially looks simple but is actually very sophisticated.

> …but the researchers are entertaining a more intriguing option, as well: that these tiny shifts, which often move the retina by only a degree or less in visual angle, improve the resolution of fly vision. Retinal movements may thus help to explain how fruit flies, which have only about 6000 photoreceptors per eye—a trifle compared to the hundreds of millions of receptors in a human eye—can still see surprisingly well.¹

One benefit of such amazing research is that

> …optical engineers have recently been able

to boost the resolution [the smallest interval measurable by optical instruments] of cameras by introducing tiny movements [of] the sensors, a parallel to insect vision that might help to improve the future design of such cameras.[1]

This has nothing to do with evolution. Humans are copying what has been painstakingly discovered in God's living creation.

Fruit fly

References

1. Maimon, G. Fruit flies move their retinas much like humans move their eyes. Rockefeller University news release. Posted on rockefeller.edu October 26, 2022, accessed October 31, 2022.
2. Sherwin, F. To Study Human Brains, Evolutionists Studied…Ants. *Creation Science Update.* Posted on ICR.org November 8, 2021, accessed October 31, 2022.
3. Fenk, L. et al. 2022. Muscles that move the retina augment compound eye vision in Drosophila. *Nature.* 612: 116–122.
4. Bethell, T. 2017. *Darwin's House of Cards.* Seattle, WA: Discovery Institute Press, 115.

Flame skimmer dragonfly

8
AERIAL ENGINEERING AND PHYSICS OF THE DRAGONFLY

Frank Sherwin, D.Sc. (Hon.)

Dragonflies (order Odonata) are perhaps one of the most studied and appreciated insects in the world today. Like the hummingbird, the dragonfly is a master in the art of flight. New research has only increased the sheer amazement at this four-winged wonder.[1]

A 2022 dragonfly investigation involved Cornell University scientists "untangl[ing] the intricate physics and neural controls that enable dragonflies to right themselves while they're falling."[2] The zoologists discovered that the chain of righting mechanisms began with the remarkable eyes of the insect that can receive 200 images per second. Dr. Jeffrey Tomkins wrote,

> The dragonfly's optics are also amazing, with almost its entire head composed of visual sensors loaded with engineering that's only beginning to be understood. It has very complex eyes constructed of individual visual sensory units called ommatidia. A single compound eye has an integrated lens system

containing up to 30,000 ommatidia. Each individual ommatidium collects its own stream of visual information that's transmitted to the dragonfly's brain, where it's decoded and processed to form a mosaic image with intricate visual depth and detail.[3]

From the information gained by its incredible optics, neural signals are generated and transmitted to the dragonfly's wings, all in a fraction of a second.

That visual cue triggers a series of reflexes that sends neural signals to the dragonfly's four wings, which are driven by a set of direct muscles that modulate the left-wing and right-wing pitch asymmetry accordingly. With three or four wing strokes, a tumbling dragonfly can roll 180 degrees and resume flying right-side up. The entire process takes about 200 milliseconds.[2]

Clearly, the engineering ability of a winged insect to control its flight in milliseconds comes from the mind of an all-wise Creator.

"Flight control on the timescale of tens or hundreds of milliseconds is difficult to engineer," [Jane Wang, professor of mechanical engineering and physics in the College of Arts & Sciences,] said. "Small flapping machines now can take off and turn, but still have trouble remaining in the air. When they tilt, it is hard to correct. One of the things that animals have to do is precisely solve these kinds of problems."[2]

Dragonfly

The various animals do not "solve these kinds of problems," of course. The Lord Jesus designed all animals with the innate ability to correct potential problems.

Is such split-second modulation that requires "complex mathematical modeling to understand the mechanics of insect flight" the result of chance and time or purpose and plan?

References

1. Wang, Z. et al. 2022. Recovery mechanisms in the dragonfly righting reflex. *Science.* 376 (6594): 754.
2. Dragonflies use vision, subtle wing control to straighten up and fly right. *ScienceDaily.* Posted on sciencedaily.com May 13, 2022, accessed May 22, 2022.
3. Tomkins, J. Intricate Animal Designs Demand a Creator. *Creation Science Update.* Posted on ICR.org June 28, 2019, accessed May 22, 2022.

9
OPEN OCEAN DRAGONFLY MIGRATION BOGGLES THE MIND

Jeffrey P. Tomkins, Ph.D.

Animal migrations occur all over the earth among many types of creatures, with some winged creatures (birds and insects) making the most extreme and lengthy ones. Among insects, the globe skimmer dragonfly (*Pantala flavescens*) is exceptional—flying up to 3,730 miles across the open ocean. Scientists are finally beginning to unravel the required specificity behind the anatomical, behavioral, and metabolic complexity that enable this amazing feat.[1]

Strong evidence implies that the extreme migration of the globe skimmer dragonfly goes across the daunting ocean expanse between the Maldives off the coast of India to East Africa. However, the small size of the creature (which is only about 1.77 inches long with a wingspan of just a little over three inches) seemed to present problems at first due to the inherent limitations of the insect's ability to store enough energy reserves. In other words, its gas tank did not appear big enough to hold the fuel needed to make the long journey.

Wandering glider
(Pantala flavescens)

In a 2021 study, a group of researchers first derived a baseline by determining the insect's specific metabolic characteristics. Then they calculated how long it could stay airborne using the maximum energy stored in its body, such as fat reserves.[1] And because other flying creatures like birds depend heavily on wind patterns, they also calculated weather models to see if available seasonal wind patterns in the migration route could facilitate the open ocean flight in both directions.

The scientists discovered a flight model that would allow successful open ocean migration—one that combined active wing flapping with gliding and that took advantage of seasonal wind patterns. In fact, there was a strong behavioral requirement for the dragonfly to select favorable wind patterns. The researchers also discovered that the specialized metabolism and physiological endurance of the dragonfly played a key role in the migration, making it all possible.

Extreme creature accomplishments, like this daunting open ocean migration across thousands of miles, and the features that allow creatures to do them boggle the human mind and utterly defy evolutionary interpretations regarding the dragonfly's origins. How could random mythical processes result in the perfect combination of anatomy, physiology, and behavioral adaptations needed for this creature to succeed in this amazing endeavor? The only logical inference we can make is that this incredible engineering was built into these dragonflies by an omnipotent Creator, the Lord Jesus Christ.

Reference

1. Hedlund, J. et al. 2021. Unraveling the World's Longest Non-stop Migration: The Indian Ocean Crossing of the Globe Skimmer Dragonfly. *Frontiers in Ecology and Evolution.* 9: 525.

10
WHY DON'T RAINDROPS BOMB BUTTERFLY WINGS?

Brian Thomas, Ph.D.

Okay, I admit most folks have probably not thought to ask this creation question. But a bigger question gets answered when we examine the fantastic butterfly features that counter the force of falling raindrops.

Butterfly wings are quite thin. How do these tiny creatures cope with raindrops that fall at 22 miles per hour? Cornell scientist Sunghwan Jung led a project that tested water drop impacts at real raindrop speeds.[1] It turns out that special surface structures on butterfly wings manage the drop impacts, which Professor Jung compared to the force of bowling balls falling from the sky on humans![2]

How do these special surfaces manage killer raindrops? At the level seen only with a microscope, we find the wings covered in rough bumps. If a drop hits flat on a sheet of glass, its force spreads in a widening wave. But when a raindrop hits a butterfly wing, the tiny bumps rupture that spreading force so that one big drop shatters into dozens of tiny droplets.

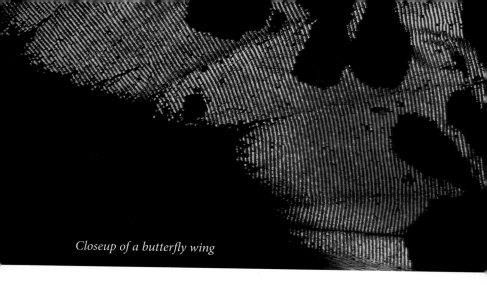

Closeup of a butterfly wing

Zoom in even closer to nanoscopic levels and we find wax structures that make the wings water-resistant. The droplets bounce right off. Without them, water would stay on the wings longer, and that would wick too much heat from tiny insect bodies. The research team found that this ingenious, thin surface cuts the water contact time by 70%.[1] The wings stay dry and whole, and the tiny flying animals stay warm and airborne.

So, these scientists found the answer to our question: butterfly wings resist raindrop power using clever micro-bumps and nano-waxes. The researchers also discovered these structures on dragonfly and moth wings, plant leaves, and even bird feathers. That leads us to the bigger question: If these amazing features help keep these creatures alive and in flight, how did such structures arise?

Many scientists attribute creature design to random changes guided by natural factors over eons. But whoever crafted these super-surfaces must have done it with flight in mind right from the start. These surfaces combine with the lightweight but strong materials and structures, navigation systems, functional landing gear, and aerodynamic shapes that all work together to make a butterfly flutter. Nature knows nothing, let alone the precise placement of parts required for flight. Truly the Lord Jesus, not natural processes, deserves all the credit for rainproofing butterfly wings before the first rain fell.

References

1. Kim, S. et al. 2020. How a raindrop gets shattered on biological surfaces. *Proceedings of the National Academy of Sciences.* 117 (25): 13901–13907.
2. Ramanujan, K. Armor on butterfly wings protects against heavy rain. *Cornell Chronicle.* Posted on cornell.edu on June 8, 2020, accessed June 15, 2020.

11
BUTTERFLY WING DESIGN REPUDIATES EVOLUTION

Jeffrey P. Tomkins, Ph.D.

The takeoff and flight of butterflies has long been derided by evolutionists as being an unstable and inefficient product of evolution. However, a 2021 study has shown that the spectacular complexity and efficiency of butterfly wing design is an optimized system worthy of emulating in a new generation of flying robotic drones.[1]

Butterflies fluttering around a sunny garden grab our curiosity and fascinate us like no other creature. In fact, butterflies look like no other flying animal because they have such unusually broad and large wings relative to their small body size. Thus, conventional scientists have been mystified as to how this specific type of flight could have come about. When evolutionary scientists first began to study butterfly wings, they claimed that they utilized unsteady aerodynamic mechanisms and that the upstroke of the wings, known as wing clap, was a particularly inefficient feature.

In this study, which was considerably more high-

Luna moth wing scales

tech than previous projects, scientists analyzed the wing action and aerodynamics of a type of butterfly called the silver-washed fritillary (*Argynnis paphia*). They used a technique called tomographic particle image velocimetry, which measures the velocity of objects in three dimensions. They also applied a mechanical analysis called kinematics, which defines the motion of an object without any reference to the forces that cause it. The data for these analyses were obtained by the high-speed filming of butterflies during take-off and flight in a specialized wind tunnel.

The results of the research were startling. Other free-flying creatures, including other types of insects, lack the unique type of mechanism observed, so the wing design of the butterfly was totally unexpected.

During an upward stroke, the optimized design of the flexible wings produces a specialized cavity that creates an air-filled pocket. As the wings continue to compress, the air is forced out like a jet engine, propelling the butterfly forward. The downward wing stroke stabilizes the flight pattern and keeps the butterfly in the air. Not only does this mechanism allow for efficient flight, but it also allows for rapid takeoff when the butterfly needs to escape a predator.

When they compared their butterfly data to that of insects with more rigid wings, the researchers were able to demonstrate an increase in forceful impulse of more than 22% combined with an increase in overall efficiency of more than 28%. One must consider that this boost in power and efficiency would also likely contribute to overall resilience. For example, the monarch butterfly makes a lengthy migration across North America where it has been shown to fly 50 to 100 miles per day.

Needless to say, this fabulous wing engineering was immediately scrutinized for its potential to benefit mankind. The authors of the paper stated, "Furthermore, our findings could aid the design of man-made flapping drones, boosting propulsive performance."[1]

However, as is the norm in conventional research publications, no credit or glory was given to our mighty Creator, the Lord Jesus Christ, whose handiwork was clearly revealed. As the Bible says,

> For since the creation of the world His invisible attributes are clearly seen, being under-

stood by the things that are made, even His eternal power and Godhead, so that they are without excuse, because, although they knew God, they did not glorify Him as God, nor were thankful, but became futile in their thoughts, and their foolish hearts were darkened.[2]

References

1. Johansson, L.C. and P. Henningsson. 2021. Butterflies fly using efficient propulsive clap mechanism owing to flexible wings. *Journal of The Royal Society Interface*. 18 (174): 20200854.
2. Romans 1:20.

12
HUMMINGBIRDS BY DESIGN
Frank Sherwin, D.Sc. (Hon.)

Zoologists have wonder and appreciation for the animals they investigate, whether the creatures fly through air, swim in water, or walk on land. Stanford University biologist Vadim Pavlov stated it best: "Animals are exciting sources of elegant engineering solutions in aero and hydrodynamics."[1]

This is certainly true regarding the hummingbird, an amazing, feathered acrobat, tiny and fearless. Many scientists attempt to explain hummingbirds' profound innate engineering in Darwinian terms, but the evolutionists' own explanations reveal the flaws in their logic.

Of all the birds of God's creation, the hummingbird ("hummer") might be the most remarkable. Everything about these creatures shout design.[2] Their little hearts can beat more than 1,000 times a minute, while their wings beat from 50 to 80 times a second. But evolutionists see hummingbirds as "diversifying and evolving over millions of years" and marvel that "evolution can take an animal to such extremes."[3]

Evolutionary theory maintains these tiny birds

evolved from a non-hummingbird ancestor. According to a study led by UC Berkeley herpetologist Jimmy McGuire:

> The branch leading to modern hummingbirds arose about 42 million years ago when they split from their sister group, the swifts and treeswifts. This probably happened in Europe or Asia, where hummingbird-like fossils have been found dating from 28-34 million years ago.[3]

But when we look at the fossil record, we find that hummingbirds have always been hummingbirds. An article in the *Journal of Ornithology* validates this:

> A near-complete, partially articulated skeleton of a hummingbird was recently found in the menilite shales of the Polish Flysh Carpathians. The specimen is dated to the Early Oligocene [about 33 million years ago]. It shares derived characters [traits absent in the last common ancestor] with extant [living] hummingbirds and plesiomorphic characters with swifts.[4]

The same article also identified the Oligocene fossil "as a new species of the same genus [*Eurotrochilus*]."[4] In other words, the fossil was still very much a hummingbird. The journal *Science* also described "tiny skeletons of stem-group hummingbirds from the early Oligocene of Germany that are of essentially modern appearance and exhibit morphological specializations toward nectarivory and hovering flight." The paper referred to these as "the oldest fossils of modern-type hummingbirds, which had not previously been reported from the Old World."[5]

Evolutionists believe the hummer somehow evolved from Archosauria (class Reptilia), the group that includes crocodiles and alligators. However, the fossil record doesn't document this at all. The alleged common ancestor of hummingbirds remains unknown.

Hummingbirds and Flowers

Many are familiar with the relationship between the hummer and the flowers they feed on and pollinate. These flowers seem uniquely designed to accommodate a hummingbird's beak and tongue, but did such a relationship evolve? A study published in *Integrative Organismal Biology* stated, "The fit between flowers and hummingbird bills has long been used as an example of impressive co-evolution."[6]

Yet coevolution is hardly an impressive explanation because it isn't scientific. Two evolutionists

describe coevolution as being "caused by selection pressures" that each species somehow enacts on the other.[7] Selection pressure is supposedly exerted by the environment. But the environment can't apply pressure. Nature can't think and select—it just exists. For example, when it comes to bird vision switching between

Golden-tailed sapphire hummingbird

violet and ultraviolet wavelengths "in the course of avian evolution," it isn't surprising that evolutionary scientists admit "the selection pressures driving this switch are not well understood."[8]

God designed the hummingbird's bill and tongue to drink nectar from flowers many insects wouldn't be able to access. Just as hummingbirds have always been hummingbirds, we find that flowers (angiosperms) have always been flowers as God created them. They didn't evolve. Two separate published studies powerfully summarized the "mystery" of the origin of flowers. One stated: "The question of the structure and biology of the ancestral angiosperms, and especially their flowers, is an enduring riddle."[9] The other study's lead author observed that "the discrepancy between estimates of flowering plant evolution from molecular data and fossil records has caused much debate."[10]

Most of us don't appreciate the complexity of what happens when a hungry hummer feeds from a flower or man-made feeder. High-speed filming and detailed anatomical studies revealed the birds are designed with a long, forked tongue that aids in drinking biomechanics. The tongue opens up when inserted into the flower, and the nectar is pumped up the tongue via two grooves. The hummer can do this up to 20 times per second.

Because of this speed, scientists realized the traditional explanation of nectar being drawn up by capillary action (the movement of a liquid in a narrow space caused by the surface tension of the liquid—adhesion and cohesion) was incorrect. It was wrongly assumed for almost two centuries that hummingbirds took in nectar by capillary rise loads. Although the hummingbird's amazing feeding mechanism has now been shown, evolutionists maintain that its tongue's abilities—no matter how complex—evolved. "Nectarivores [nectar-eating organisms], however, have evolved specialized tongues that function as their primary food-capturing device."[11] But "have evolved" isn't a scientific explanation. Creationists assert that hummers, like woodpeckers, were designed with specialized tongues from the beginning.

Hummingbird feeding appears to be irreducibly complex—a phrase non-evolutionists use to explain the way in which a number of crucial parts must all work together for a structure or process to function. In this case, it starts with the tongue fitting the flower. According to a *New York Times* report on hummingbird tongue research:

> The findings could affect thinking about how flowers and hummingbirds have evolved together, since the shape of the flower, the composition of the nectar and the shape and workings of the tongue *must all fit together for the system to work.*[12]

Nectar taken into the hummingbird's body is immediately metabolized (burned) for energy to power rapid wing strokes. Sugars are compounds rich in energy. Evolutionists say, "Whereas humans evolved over time on a complex diet, hummingbirds evolved on a diet rich in sugar."[13] But hummingbirds and people haven't evolved from ancient bird and human ancestors. It's far more accurate to say humans are *designed* to metabolize a complex diet, whereas hummingbirds are *designed* to metabolize a diet rich in sugar.

Because of their immediate energy needs, God designed hummingbirds to process the monosaccharides fructose and glucose with ease. When chemically hooked together, these make up disaccharide sucrose, or table sugar. University of Toronto biologist Kenneth Welch stated,

> What's very surprising is that unlike mammals such as humans, who can't rely on fructose to power much of their exercise metabolism, hummingbirds use it very well. In fact,

Black jacobins hummingbird (Florisuga fusca)

they are very happy using it and can use it just as well as glucose.[13]

Hummingbird Vision

Hummingbird vision is also a marvel. The Creator designed hummingbirds with a color vision range that exceeds that of humans. This unique vision helps birds see nectar-producing plants, potential mates, predators, and objects within their range.

Recent research verifies this amazing fact. In addition to the three types of color cones humans have in their eyes, birds have one more. Not only are they sensitive to red, green, and blue light, they also can pick up ultraviolet rays.[14] In a Princeton University news release, evolutionary biologist Mary Caswell Stoddard said,

> Not only does having a fourth color cone type extend the range of bird-visible colors into the UV [ultraviolet], it potentially allows birds to perceive combination colors like ultraviolet+green and ultraviolet+red—but this has been hard to test.[14]

The release went on to state, "Stoddard and her colleagues designed a series of experiments to test whether hummingbirds can see these nonspectral colors," the results of which were published in the *Proceedings of the National Academy of Sciences*.[14] And indeed, the hummingbird's fourth color cone type allows it to see colors we cannot even imagine! In their study, Stoddard's team reported:

> Nonspectral colors are perceived when non-

adjacent cone types (sensitive to widely separated parts of the light spectrum) are predominantly stimulated. For humans, purple (stimulation of blue- and red-sensitive cones) is a nonspectral color; birds' fourth color cone type creates many more possibilities.[15]

An earlier study conducted by the Washington University School of Medicine in St. Louis reported:

> The researchers characterized the carotenoid pigments from birds with violet vision and from those with UV vision and used computational models to see how the pigments affect the number of colors they can see.... The study also revealed that sensitivity of the violet/UV cone and the blue cone in birds must move in sync to allow for optimum vision. Among bird species, there is a strong relationship between the light sensitivity of opsins [a protein contained in vertebrates' visual pigment that determines the pigment's spectral sensitivity] within the violet/UV cone and mechanisms within the blue cone, which coordinate to ensure even UV vision.[16]

Was such intricate design of these cones caused by chance and time or plan and purpose?

Hummingbird Communication

Hummingbirds vocalize using high-pitched chirping. "Hummingbirds are a fascinating group of birds, but some aspects of their biology are poorly understood, such as their highly diverse vocal behaviors."[17] Like their vision, these birds' vocalization is incredibly complex.

> The predominant vocalization of black jacobins (*Florisuga fusca*) [a type of hummingbird]...consists of a triplet of syllables with high fundamental frequency (mean F0 ~11.8 kHz), rapid frequency oscillations and strong ultrasonic harmonics and no detectable elements below ~10 kHz. These are the most common vocalizations of these birds, and their frequency range is above the known hearing range of any bird species recorded to date, including hearing specialists such as owls. These observations suggest that black jacobins either have an atypically high frequency hearing range, or alternatively their primary vocalization has a yet unknown function unrelated to vocal communication. Black jacobin vocalizations challenge current notions about vocal communication in birds.[17]

Hummingbird Physiology

Animals can make physiological adjustments to slow down their metabolism, and this can affect body temperature. For example, hummingbirds do this on cold nights to save energy. A recent investigation added another layer to the astounding feats of these tiny marvels:

> Researchers sought to learn more about several of the species of hummingbirds that live in the Andes in South America—a region that can get very cold at night in the summer.

In this case, the researchers looked at species that survive at altitudes up to 3,800 meters above sea level....The researchers also found that one of the birds, a black metal tail, lowered its body temperature to just a few degrees above freezing—to 3.3 degrees C. This finding marked a record low body temperature for any non-hibernating mammal.[18]

The wonderful, created hummingbird is a miracle. Its specialized interaction with plants, visual color range, complex vocalization, and unique flight and physiological capabilities are testaments to God's living creation design.

References

1. Kubota, T. Stanford researchers discover biological hydraulic system in tuna fins. *Stanford News Service.* Posted on news.stanford.edu July 20, 2017, accessed September 8, 2020.
2. Thomas, B. 2016. Hummingbirds! *Acts & Facts.* 45 (4): 16.
3. Michelson, M. Hummingbird Evolution. California Academy of Sciences. Posted on calacademy.org April 9, 2014, accessed September 8, 2020.
4. Bochenski, Z. and Z. Bochenski. 2008. An Old World hummingbird from the Oligocene: A new fossil from Polish Carpathians. *Journal of Ornithology.* 149 (2): 211–216.
5. Mayr, G. 2004. Old World Fossil Record of Modern-Type Hummingbirds. *Science.* 304 (5672): 861–864.
6. Rico-Guevara, A. et al. 2019. Shifting Paradigms in the Mechanics of Nectar Extraction and Hummingbird Bill Morphology. *Integrative Organismal Biology.* 1 (1): oby06.
7. Thain, M. and M. Hickman. 2004. *The Penguin Dictionary of Biology.* New York: Penguin Books, 161.
8. Toomey, M. B. et al. Complementary shifts in photoreceptor spectral tuning unlock the full adaptive potential of ultraviolet vision in birds. *eLife.* Posted on elifesciences.org July 12, 2016, accessed September 8, 2020.
9. Endress, P. K. and J. A. Doyle. 2009. Reconstructing the ancestral angiosperm flower and its initial specializations. *American Journal of Botany.* 96 (1): 22–66.

10. When did flowers originate? University College London news release. Posted on ucl.ac.uk February 5, 2018, accessed September 8, 2020.
11. Rico-Guevara, A. and M. A. Rubega. 2011. The hummingbird tongue is a fluid trap, not a capillary tube. *Proceedings of the National Academy of Sciences.* 108 (23): 9356–9360.
12. Gorman, J. The Hummingbird's Tongue: How It Works. *The New York Times.* Posted on nytimes.com September 8, 2015, accessed September 14, 2020. Emphasis added.
13. Hummingbird metabolism unique in burning glucose and fructose equally. University of Toronto media release. Posted on media.utoronto.ca December 5, 2013, accessed September 8, 2020.
14. Fuller-Wright, L. Wild hummingbirds see a broad range of colors humans can only imagine. Princeton University news release. Posted on princeton.edu June 15, 2020, accessed September 8, 2020.
15. Stoddard, M. C. et al. 2020. Wild hummingbirds discriminate nonspectral colors. *Proceedings of the National Academy of Sciences.* 117 (26): 15112–15122.
16. How birds unlock their super-sense, ultraviolet vision. *eLife.* Posted on elifesciences.org July 12, 2016, accessed September 8, 2020.
17. Olson, C. R. et al. 2018. Black Jacobin hummingbirds vocalize above the known hearing range of birds. *Current Biology.* 28 (5): R204–R205.
18. Yirka, B. Hummingbird reduces its body temperature during nightly torpor. *Phys.org.* Posted on phys.org September 9, 2020, accessed September 10, 2020.

Anna's hummingbird

13
HUMMINGBIRD FLIGHT STRATEGIES
Frank Sherwin, D.Sc. (Hon.)

The fossil record shows that hummingbirds have always been hummingbirds. Such amazing mastery of aerial acrobatics could never have evolved.[1,2] ICR's Dr. Jeff Tomkins put it succinctly:

> The hummingbird is another animal that glorifies the Creator. This little creature is distinctly different from all other bird kinds. Hummingbirds are the only bird that can fly backward. They can literally zip around in just about every direction due to their wings' ability to rotate in a full circle and flap up to 80 times per second.[3]

Thus, hummingbird flight is not only amazing, but it is also unique among all animals. These specialists, which include the smallest bird in the world, can navigate efficiently with their long, saber-like wings. In fact, "the fast turns of hummingbirds have inspired a massive motion-tracking study of the birds' flight maneuvers," stated science writer Susan Milius of *Science News*.[4] She went on to report,

> Bigger hummingbird species don't seem

handicapped by their size when it comes to agility. A battleship may not be as maneuverable as a kayak, but in a study of 25 species, larger hummingbirds outdid smaller species at revving or braking while turning. Measurements revealed these species have more muscle capacity and their wings tended to be proportionately larger for their body size than smaller species. Those boosts could help explain how these species could be so agile despite their size, researchers report...[4]

Additional investigations have recently revealed that an Anna's hummingbird is able to navigate effortlessly and swiftly through an intricate maze of complex environments with bilaterally asymmetric wing motions. Marc Badger and other scientists writing in the *Journal of Experimental Biology* described how this is accomplished:

> Many birds routinely fly fast through dense vegetation characterized by variably sized structures and voids. Successfully negotiating these cluttered environments requires maneuvering through narrow constrictions between obstacles. We show that Anna's hummingbirds (*Calypte anna*) can negotiate apertures less than one wingspan in diameter using a novel sideways maneuver that incorporates continuous, bilaterally asymmetric wing motions. Crucially, this maneuver allows hummingbirds to continue flapping as they negotiate the constriction. Even smaller openings are negotiated via a faster ballistic

trajectory characterized by tucked and thus non-flapping wings..."[5]

Blind evolutionary forces like random mutations could never perfect the hummer's fast ballistic trajectory, obstacle avoidance, and novel sideways maneuvers we observe today. Instead, we clearly see evidence of purpose, plan, and special creation in these highly specialized creatures.

What can scientists learn from observing the detailed and precise aerobatics of hummingbirds? Five insightful researchers pointed to engineering applications when they stated that "These strategies for aperture transit and associated flight trajectories can inform designs and algorithms for small aerial vehicles flying within cluttered environments."[5]

Yes, scientists actually learn from what our Creator, the Lord Jesus, designed just thousands of years

Anna's hummingbird

ago. We should always be ready to give glory to the Creator and not to the creation, as Paul stated in the book of Romans.[6]

References

1. Sherwin, F. 2020. Hummingbirds by Design. *Acts & Facts.* 49 (11): 17–19.
2. Thomas, B. 2016. Hummingbirds! *Acts & Facts.* 45 (4): 16.
3. Tomkins, J. 2019. Intricate Animal Designs Demand a Creator. *Acts & Facts.* 48 (7): 14.
4. Milius, S. Trove of hummingbird flight data reveals secrets of nimble flying. *Science News.* Posted on sciencenews.org February 18, 2018, accessed November 20, 2023.
5. Badger, M. et al. 2023. Sideways maneuvers enable narrow aperture negotiation by free-flying hummingbirds. *Journal of Experimental Biology.* 226 (21).
6. Romans 1:25.

14
THE SYRINX SONG
Frank Sherwin, D.Sc. (Hon.)

The rippling murmur of a mountain brook, the intertwining notes of a Chopin nocturne, and the melodious sounds of most birds are a tonic to soothe the soul.

What makes the unique bird sounds is a structure called the syrinx, found at a point where the trachea, or windpipe, splits into the bronchi, the passageways to the lungs. The syrinx is typically designed with a resonating chamber and elastic vibrating connective tissues called tympaniform membranes. Sound is produced when the membranes are pushed inward via muscular contraction so they partially block the bronchi.

It's been known for centuries that some songbirds can make more than one sound at a time, but how? Through painstaking research (such as endoscopic techniques and high-resolution 3-D images), ornithologists were able to determine some birds can control the lateral and medial labia of the syrinx and produce an amazing effect called lateralization. "Each side of the syrinx receives its own motor program

Chaffinch singing

that, together with that sent to respiratory muscles, determines the acoustic properties of the ipsilaterally produced sound."[1]

Did such sub-millisecond precision come about by time and chance or by plan and purpose? Indeed, "scientists aren't sure how or why birds evolved these unique voiceboxes,"[2] and "why only birds evolved a novel sound source at this location remains unknown, and hypotheses about its origin are largely untested."[3]

Even with 21st-century technology, "the anatomy of the complex physical structures that generate [a bird's] sound have been less well understood."[4] In addition, there appears to be no syrinx evolution. The first time a syrinx is found in the fossil record, it's 100% a syrinx.[5]

Recent conventional explanations for syrinx evolution are anemic. "The longer trachea of birds compared to other tetrapods made them likely predisposed for the evolution of a syrinx."[3] The phrase "likely predisposed" isn't a scientific explanation. The same publication appeals to "strong selective pressures," which is also hardly scientific.

Evolutionist Chad Eliason of Chicago's Field Museum, who is committed to the strange idea that heavy-tailed theropod dinosaurs somehow became hummingbirds, stated, "If we found fossil evidence of a syrinx in dinosaurs, that would be a smoking gun, but we haven't yet. In the meantime, we have to look at other animals for clues."[2]

Researchers did look at other animals but to no avail.

In a new study in the *Proceedings of the Natural Academy of Science,* an interdisciplinary team of developmental biologists, evolutionary morphologists, and physiologists examined the windpipes of birds, crocodiles, salamanders, mice, and cats to learn more about how syrinxes evolved. Their findings seem to confirm: the syrinx is an evolutionary odd duck.[2]

Why did God place the syrinx where He did in birds? A team of evolutionists explain why without giving Him the credit: "By sitting so low in the airway, the syrinx can produce sound with very high efficiency."[6]

The syrinx is a uniquely designed and incredibly complex organ the first time it's found in the fossil record. It has no evolutionary history. It's designed to work at superfast speeds to produce some of the most beautiful music in God's creation. Conventional scientists might consider it an odd duck, but creationists know it is a marvelous manifestation of its Maker's ingenuity.

References

1. Suthers, R. A. 1997. Peripheral control and lateralization of birdsong. *Journal of Neurobiology.* 33 (5): 632–652.
2. Birds' voiceboxes are odd ducks. *ScienceDaily.* Posted on sciencedaily.com September 24, 2018, accessed May 28, 2019.
3. Riede, T. et al. 2019. The evolution of the syrinx: An acoustic theory. *PLOS Biology.* 17 (2): e2006507.

4. How do songbirds sing? In 3-D! *ScienceDaily*. Posted on sciencedaily.com January 8, 2013, accessed May 28, 2019.
5. Clarke, J. A. et al. 2016. Fossil evidence of the avian vocal organ from the Mesozoic. *Nature*. 538 (7626): 502–505.
6. Birds' Surprising Sound Source: The best place for a bird's voice box is low in the airway, researchers find. University of Utah news release. Posted on unews.utah.edu April 10, 2019.

15
CREATED CUTTLEBONE'S SWEET SPOT
Frank Sherwin, D.Sc. (Hon.)

God created Earth during the creation week just thousands of years ago. This includes its marvelous oceans—128 million square miles of salt water. In the 21st century, scientists and laymen alike are struck with the amazing variety of oceanic creatures found at all depths. Some of the most fascinating marine animals are the cephalopods, which include squid, octopus, and cuttlefish. God designed the cuttlefish (family Sepiidae) with a strange and important structure called the cuttlebone, a brittle, internal shell designed with gas-filled chambers that's used for buoyancy control.

You've probably seen cuttlebone without recognizing what it was. It's used as a dietary supplement for pet birds, placed in their cages as an important source of calcium. Zoologists have recently been fascinated by the design of cuttlebone at the microscopic level. The bone is not exactly robust, yet it can tolerate a great deal of damage. A recent Virginia Tech news release spotlighted a study led by mechanical engineering professor Ling Li, who heads the Laboratory for Biological and Bio-inspired Materials.[1] The

Common cuttlefish

article stated, "The more Li studies these animals, the more he's amazed by the uses their bodies find for intrinsically brittle and fragile material. Especially when the use defies that fragility."[2]

The design inference of cuttlebone and other biological materials is undeniable.[3] Researchers "found that the microstructure's unique, chambered 'wall-septa' design optimizes cuttlebone to be extremely lightweight, stiff, and damage-tolerant."[2] They found that under various magnification techniques—such as powerful X-ray beams—the shell's microstructure is made up of "wavy walls instead of straight struts. The waviness increases along each wall from floor to ceiling in a 'waviness gradient.'"[2]

> [The] cuttlebone's wavy walls induce or control fractures to form at the middle of walls, rather than at floors or ceilings, which would cause the entire structure to collapse. As one chamber undergoes wall fracture and subsequent densification—in which the fractured walls gradually compact in the damaged chamber—the adjacent chamber remains intact until fractured pieces penetrate its floors and ceilings.[2]

This wall-septa design gives cuttlebone control of where and how damage occurs in the shell. It allows for graceful, rather than catastrophic, failure: when compressed, chambers fail one by one, progressively rather than instantaneously.[2]

Not surprisingly, evolution is not mentioned in this article. Why should it be? The cuttlebone microstructure speaks of intricate design, specific purpose, and deliberate plan—denying random processes. The Creator designed it to have an ideal point between the absorption of energy and toughness. The researchers recognized this:

> We show that cuttlebone sits in an optimal spot. If the waviness becomes too big, the structure is less stiff. If the waves become smaller, the structure becomes more brittle. Cuttlebone seems to have found a sweet spot, to balance the stiffness and energy absorption.[2]

Did the cuttlebone "find" this sweet spot? No,

the Creator put it there. The Master Engineer is to be praised for such sophistication in His living world.[4]

References

1. Yang, T. et al. 2020. Mechanical design of the highly porous cuttlebone: A bioceramic hard buoyancy tank for cuttlefish. *Proceedings of the National Academy of Sciences*. 117 (38): 23450–23459.
2. Researchers find cuttlebone's microstructure sits at a 'sweet spot' for lightweight, stiff, and damage-tolerant design. Virginia Tech Daily. Posted on vtnews.vt.edu September 11, 2020.
3. Sherwin, F. 2002. God's Creation Is 'Clearly Seen' in Biomechanics. *Acts & Facts*. 31 (3).
4. Sherwin, F. 2017. Architecture and Engineering in Created Creatures. *Acts & Facts*. 46 (10): 10–12.

16
MOLECULAR MOTORS OF A SQUID SHOW CET IN ACTION

Frank Sherwin, D.Sc. (Hon.)

It has traditionally been thought in biology that invertebrates were somehow simple and less complex than vertebrates. But in past decades, this has been turned on its head.[1,2] For example, Cephalopods (marine molluscs) continue to amaze researchers with their intellect and unparalleled complexity.[3] This is especially true for the genetics and biochemistry that regulate such complexity at the cellular level.

Microtubules (MTs) are extremely tiny, tube-like structures found in the cells of people, plants, and animals. They contribute to cell shape (cytoskeleton), cell division (mitosis), and a flexible array of scaffolds upon which organelles and other intracellular components are transported by motor proteins. Two-headed motor proteins called kinesin and dynein actually walk in a coordinated fashion along these numerous submicroscopic microtubules.[4]

In cephalopods, such as squid and octopus, the expression patterns of many transcripts (an RNA copy of a piece of DNA) are modified by RNA editing.

Inshore squid

RNA editing is a widespread epigenetic process that can alter the amino acid sequence of proteins, termed "recoding." In cephalopods, most transcripts are recoded, and recoding is hypothesized to be an adaptive strategy to generate phenotypic plasticity. However, how animals use RNA recoding dynamically is largely unexplored. We investigated the function of cephalopod RNA recoding in the microtubule motor proteins kinesin and dynein.[5]

Epigenetics is the study of changes in organisms caused by modification of gene expression rather than alteration of the genetic code itself.[6,7]

More remarkably, report two scientists at University of California San Diego in a new study, at least some cephalopods possess the ability to recode protein motors within cells to adapt to different water temperatures….

"This work supports the idea that recoding in cephalopods is important for dynamically tuning protein function to support physiological needs and acclimate to changing environmental conditions," said Reck-Peterson [Ph.D., a professor in the departments of Cellular and Molecular Medicine at UC San Diego School of Medicine and Cell and Developmental Biology at UC San Diego]. "These animals are taking a completely unique approach to adapting to their surroundings."[8]

In other words, these amazing inshore squid (*Doryteuthis opalescens*) function in line with the ICR model of CET: a process of continuous environmental tracking that generates adaptational changes (traits that are modified through development, biochemistry, physiology, or the expression of particular groups of genes) for creatures in different environments. In fact, the article states, "the squid 'editome' may be a valuable resource for highlighting regions of molecules that are amenable to plasticity or change."[8]

Indeed, one would be hard-pressed to find a better example of CET. Natural selection is not mentioned in this report. There's no need. The squid have been created with the genetic ability to move in and fill various aquatic niches. They must "have at least three essential elements: *a sensor* to detect changing environmental conditions, a *logic mechanism* to select suitable responses, and *actuators* to implement those responses."[7] This is an example of a creature that is thoughtfully engineered with adaptative programming built into it by the Creator, the Lord Jesus. Such designed adaptive engineering enables the squid to continuously track and appropriately self-adjust to specific environmental changes, including a broad range of ocean temperatures.

"The work suggests that squid can tune their

proteome (an organism's entire complement of proteins) on the fly in response to changes in ocean temperature," said Reck-Peterson. "One can speculate that this allows these marine ectotherms—animals that depend on external sources of body heat—to survive and thrive in a broad range of ocean temperatures."[8]

But they can only survive and thrive if they have these designs and adaptations built in from the beginning. Such engineering features did not come about by chance and time but from the hand of the Creator, "Which doeth great things past finding out; yea, and wonders without number."[9]

References

1. Sherwin, F. 2011. "Relatively Simple." *Acts & Facts.* 40 (7): 17.
2. Sherwin, F. Bee Brains Aren't Pea Brains. *Creation Science Update.* Posted on ICR.org July 11, 2019, accessed June 11, 2023.
3. Thomas, B. Where Did the Mimic Octopus Get Its Amazing Abilities? *Creation Science Update.* Posted on ICR.org September 14, 2010, accessed April 26, 2023.
4. Sherwin, F. Muscle Motion Discoveries Challenge Evolutionism. *Creation Science Update.* Posted on ICR.org February 6, 2013, accessed June 12, 2023.
5. Rangan, K. and S. L. Reck-Peterson. 2023. RNA recoding in cephalopods tailors microtubule motor protein function. *Cell.* 186 (12): 2531–2543.
6. Tomkins, J. Epigenetic Code More Complicated than Previously Thought. *Creation Science Update.* Posted on ICR.org January 28, 2016, accessed June 12, 2023.
7. Randy J. Guliuzza, 2018. Epigenetics…Engineered Phenotypic "Flexing". *Acts & Facts.* 47 (1): 17–19.
8. Science Writer. When water temperatures change, the molecular motors of cephalopods do too. *Phys.org.* Posted on phys.org June 8, 2023, accessed June 11, 2023.
9. Job 9:10.

Hexactinellid sponge

17
MARINE SPONGES INSPIRE

Frank Sherwin, D.Sc. (Hon.)

The oceans are alive with God's diverse and amazing creatures. Scripture tells us, "God created great sea creatures and every living thing that moves, with which the waters abounded, according to their kind."[1] Both vertebrates (e.g., sharks and whales) and invertebrates (e.g., clams and crabs) were created just thousands of years ago, including the allegedly "simple" sponge.[2]

Zoologists see "the exterior simplicity of a sponge mask[ing] chemical and functional sophistication."[3] For example, the spicules of a certain sponge (class Hexactinellida) are composed of calcareous or siliceous material designed by the Creator to transmit light via fiber optics deep into the sponge's photosynthetic tissue.

The fiber optics of siliceous spicules have now been confirmed. This has sparked interest among materials scientists and engineers in the enzymatic machinery needed to form silica nanoparticles and to fuse these particles into spicules inside and outside the sponge cells.[3]

Did this sponge achieve such sophistication by just chance and many millions of years? What was the origin of sponges—did they evolve from a non-sponge ancestor? Evolutionists can only say sponges have existed as sponges for nearly a half-billion years (the early Cambrian period) and, "according to some claims, the Precambrian."[3]

Regardless, sponge construction continues to amaze. Recently, it was reported that scientists "are using the glassy skeletons of marine sponges as inspiration for the next generation of stronger and taller buildings, longer bridges, and lighter spacecraft."[4] Why? It's because in one case,

> a deep-water marine sponge [*Euplectella aspergillum*, common name—Venus' flower basket], has a higher strength-to-weight ratio than the traditional lattice designs that have been used for centuries in the construction of buildings and bridges. "We found that the sponge's diagonal reinforcement strategy achieves the highest buckling resistance for a given amount of material, which means that we can build stronger and more resilient structures by intelligently rearranging existing material within the structure," said Matheus Fernandes, a graduate student at SEAS [Harvard John A. Paulson School of Engineering and Applied Sciences] and first author of [a related paper published in *Nature Materials*].[4]

Science writer Bruce Fellman stated, "Biomechanics studies how the design and construction of plants and animals obey and even capitalize on the laws of physics."[5] Such overt biomechanical design, as seen in the Venus' flower basket, has been discussed by the Institute for Creation Research in the past.[6] The more one studies *Euplectella* (not to mention other creatures), the more one logically comes to a design inference. "To support its tubular body, *Euplectella aspergillum* employs two sets of parallel diagonal skeletal struts, which intersect over and are fused to an underlying square grid, to form a robust checkerboard-like pattern."[4]

This is creation morphology—the bringing together of structural information as we observe, measure, and research God's creatures using the perspectives of function, form, ecology, and design. It is

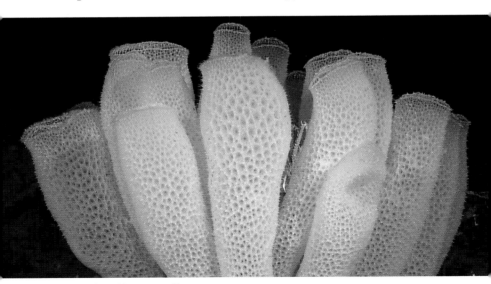

Euplectella aspergillum

perfectly natural to ascribe this living architecture to the just, loving, universal Architect of the Bible.

References

1. Genesis 1:21.
2. Sherwin, F. 2011. Relatively Simple. *Acts & Facts.* 40 (7): 17.
3. Hickman, C. et al. 2020. *Integrated Principles of Zoology,* 18th ed. New York: McGraw Hill, 253.
4. Burrows, L. Marine sponges inspire the next generation of skyscrapers and bridges. Harvard John A. Paulson School of Engineering and Applied Sciences news release. Posted on seas.harvard.edu September 21, 2020, reporting on Fernandes, M. C. et al. 2021. Mechanically robust lattices inspired by deep-sea glass sponges. *Nature Materials.* 20: 237–241.
5. Fellman, B. 1992. The Wonders of Biomechanics. *Funk & Wagnalls 1991 Science Yearbook.* New York: Funk & Wagnalls, 85.
6. Sherwin, F. 2017. Architecture and Engineering in Created Creatures. *Acts & Facts.* 46 (10): 10–12.

18
CLEVER CONSTRUCTION IN RORQUAL WHALES

Brian Thomas, Ph.D.

A few years ago, scientists discovered a unique sensory organ in the jaw of a rorqual whale—the world's largest creature. Rorqual whales, which include the blue whale and fin whale, feed by ballooning out folds of tissue that bag gobs of krill from fertile ocean waters. Some of those researchers recently described the unique bungee-cord-like nerve construction that illustrates clever and intentional design.

In a 2012 *Nature* report, researchers portrayed the sensory organ as a necessary component among a long list of precisely formed body parts required to make the rorqual whales' lunge feeding mechanism possible.[1] These include the comb-like baleen to filter out food, expandable "ventral groove blubber" with cartilaginous support bars that open like a Chinese fan, a newly discovered sensory organ, a split jaw that is loosely connected to the skull, and vibrissae (long stiff hairs) along the chin to sense prey. The whale's sensory organ detects pressures that its lower jaws endure when taking on so much water in its mouth.

Blue whale

Without this key sensor, the animal's jaw could rip apart.

Arctic and Antarctic oceans supply some of the best feeding areas for rorqual species. How do these enormous sea creatures keep from freezing as they engulf hundreds of gallons of cold ocean water? A countercurrent heat exchange arrangement of blood vessels throughout their enormous tongues protects their core body temperature. The *Science* authors who described this phenomenal design back in 1997 wrote, "All baleen whales possess countercurrent heat exchangers in their oral cavities, a physiological prerequisite allowing these endotherms to exploit the rich marine productivity of cold waters."[2]

What if all these whale feeding features were in place, but the creature had standard nerve packing? If this were the case, then when the whale's mouth ballooned and its tongue extended, the nerve would break, severing the vital pressure signal from that special sensory organ.

Fortunately, this does not happen. It looks like someone was looking out for whale well-being by providing a sheath within which these critical nerves can unfold and providing elastic tissues that retract the nerves. The journal *Current Biology* recently published a description of these newfound bungee-cord-like nerves, calling them "an essential component" of rorqual whales.[3]

Michael J. Benton, who is one of the world's foremost experts on the evolution of vertebrates, is a fellow of the Royal Society and studies at the University of Bristol. In his authoritative book *Vertebrate Paleontology* he describes whale origins, saying, "Looking at the great blue whale, 30m [98 feet] long, or a fast-swimming dolphin, it is hard to imagine how they evolved from terrestrial mammal ancestors, and yet that is what happened."[4]

The reason this is so difficult to imagine is because it never really did happen. Natural causes have not and do not craft collections of interdependent, complicated mechanisms that coordinate vital body systems. These kinds of elegant mechanisms have to come into existence all at once, not bit-by-bit. The evidence indicates creation by design, not gradual evolution by happenstance.

References

1. Pyenson, N. D. et al. 2012. Discovery of a sensory organ that coordinates lunge feeding in rorqual whales. *Nature*. 485 (7399): 498–501.
2. Heyning, J. E. and J. G. Mead. 1997. Thermoregulation in the Mouths of Feeding Gray Whales. *Science*. 297 (5340): 1128–1140.
3. Vogl, A. W. et al. 2015. Stretchy nerves are an essential component of the extreme feeding mechanism of rorqual whales. *Current Biology*. 25 (9): R360–R361.
4. Benton, M. J. 2005. *Vertebrate Paleontology*. Malden, MA: Blackwell Science, 342.

Brittle star

19
BRITTLE STARS SEE WITH THEIR SKIN

Frank Sherwin, D.Sc. (Hon.)

Echinoderms, "spiny-skinned" invertebrates, are first found in the Cambrian sedimentary rock layers as 100% echinoderms. One of the more interesting is the brittle star *Ophiocoma wendtii*. The first brittle stars (class Ophiuroidea) were found in the Early Ordovician sediments and don't show any evolution.

While snorkeling the waters of Hawaii several years ago, I found a brittle star among some rocks. After examination, I released it to slowly sink back to its home in the rocks—but instead it was immediately intercepted and devoured by a hungry fish!

Animal eyes are indeed amazing, and conventional scientists are unable to describe the origin of eyes from a supposed ancestral patch of cells.[1,2] Scientists focused research on this specific brittle star three decades ago. They were fascinated to discover that it had strong light-responding behaviors despite having no eyes.[3] The scientists found these amazing creatures had "an enormous number of microscopic crystal bumps [tiny crystal balls]" that they thought might act as "microlenses."

Brittle star

However, the researchers later found the entire body of the brittle star was covered in light-sensitive cells embedded in the skin, resulting in a huge light-sensing network. Such awesome design gives *Ophiocoma* the ability to detect shadows from a distance by seeing...*with their skin.*

The moment-by-moment information they gather from their sophisticated skin sensors is somehow utilized, but how they do this remains "a mystery."[3] One scientist exclaimed, "It's amazing that we've had this lens-based hypothesis for several decades, yet it seems that the actual mechanism behind *Ophiocoma wendtii*'s incredible behaviour is even stranger than we thought...."[3] The author of the *Phys.org* article agreed, addressing "*O. wendtii*'s incredible abilities."[3] Creation scientists see this visual system as a result of direct plan and purpose by the Creator.

They get support for their contention from an unlikely source. "'The tiny crystal balls were too similar to lenses to have been formed by chance,' says Joanna Aizenberg, of Bell Laboratories and Lucent Technologies."[4] The only alternative to evolution's long time and random chance story is the aforementioned purpose and plan. Is it any wonder this *Nature* article stated, "'It's astonishing that this organic creature can manipulate inorganic matter with such precision—and yet it's got no brain,' says Roy Sambles, who works on optics and photonics at the University of Exeter in Britain."[4]

Non-evolutionary zoologists agree with evolutionists, the features and design of *Ophiocoma*'s "visual skin" is not only strange and mysterious but also "astonishing," "incredible." But creationists give glory to the Creator, not to the creation (Romans 1:25).

References

1. Thomas, B. 2013. Amazing Animal Eyes. *Acts & Facts*. 42 (9): 16.
2. Sherwin, F. 2017. Do "Simple" Eyes Reflect Evolution? *Acts & Facts*. 46 (9): 20.
3. University of Oxford. Star-gazing on the reef: First evidence that brittle stars may 'see' with their skin. *Phys.org*. Posted on phys.org on Jan. 24, 2018, accessed on December 15, 2019.
4. Whitfield, J. 2001. Eyes in their stars. *Nature*.

Harbor Seal

20
COMMON SEALS DISPLAY EXTRAORDINARY BIOENGINEERING

James J. S. Johnson, J.D., Th.D.

Fishermen and sailors have many occasions to see wonders of the oceans.[1] All marine creatures witness to God's glory and providence, showcasing the amazing Creator He is. One such example is the common seal, also called the harbor seal.[2]

And thankfully—according to a study in Svalbard, Norway—at least one pollution problem is improving for the harbor seal, as well as for white whales and other oceanic marine mammals.

A 2020 toxicological study, published in the science journal *Environmental Pollution*, shows how certain now-banned pollutants are appearing in reduced quantities in marine mammals compared to earlier years, indicating a decrease of those pollutants in ocean waters and oceanic food chains.[3]

Fixing problems is a lot harder than causing them. This is true in land-based healthcare and in oceanic "public health" matters, such as ocean pollutants that routinely poison food chains for both people and animals.

In particular, the harbor seal and other marine mammals are slowly recovering from years of exposure to ocean-dumped pollutants called perfluoroalkyl substances.[3] The harbor seal has many other challenges in life, but at least the problem of perfluoroalkyl substance pollutants is reducing. So, that was good news to hear in 2020.

But just how special is a harbor seal anyway? This specific seal's other common name is rather prosaic, called the "common seal" in Great Britain. Yet the bioengineering wonders that God installed within this ocean-going pinniped are far from common. Even Vikings noted their swimming skills in ancient sagas.[4]

Harbor seals are the world's most common temperate-water seals, but they can tolerate a wide range of temperatures—warm, lukewarm, cool, and super-cold. They typically live along the temperate water coastlines and continental shelves of the Northern Hemisphere. But they also live in both the northern Atlantic and northern Pacific Oceans, in coastal estuaries, and sometimes even as far south as Portugal. Oddly, one subspecies only lives in freshwater.

As a marine mammal, the harbor seal is a pinniped—it has fins, not feet. It is earless and carnivorous. Being mammals, the mothers breastfeed their pups. Regarding size, the adult harbor seal ranges from five to six feet long. Males are a bit larger than females. Body weight can approach 375 pounds!

Although the harbor seal's four flippers are ideal for swimming, they are not designed for a lot of walking. So, seals move on land by undulating, like a caterpillar. These flippers have webbed digits—like fingers or toes blended together—that can be used to scratch, groom, or provide defensive movements. Yet they can stroke powerfully for precision swimming.

Amazingly, seal reproduction occurs at sea. Like humans, seal gestation lasts for about nine months. After birth, lactation lasts for about four to six weeks. They are born around 35 pounds, but their weight doubles by the time they are weaned off their mother's fat-rich milk. Within hours of being born, the seal pups can dive and swim—and their future lives will continue that habit for years to come.

Although harbor seals sometimes sleep on land, they can also sleep in the water (like whales), subconsciously surfacing for air as needed. Diving and swimming underwater are a seal's quintessential element—whether that be in the ocean, an estuarial bay (harbor), or some freshwater river flowing into the sea.

These seals can dive more than a half-mile deep when searching for food and can remain underwater for about 40 minutes (though most dives last only around five minutes). When seals dive, their God-given, interactive sensor systems (which measure oxygen levels) and pre-programmed instincts adjust their physiology to their underwater diving needs.

When the seal's face is submerged, it automatically holds its breath, its heartbeat slows by up to 90% and its blood circulation is reduced, except to the most vital organs, the heart and brain...

The dive reflex is responsible for the seal's ability to remain submerged for long periods. The harbor seal breathes out before diving, reducing its buoyancy. Also, the harbor seal has a very high blood to volume ratio, about 1.5 times that of a human. This allows a large amount of oxygen to be carried in the bloodstream instead of the lungs. The harbor seal has high myoglobin levels, allowing high levels of oxygen to be carried in the bloodstream and tissues, about 2.5 times that of a human.[5]

Even the seal's whiskers help. The nerves inside seal whiskers help sense underwater pressure changes. The whisker nerves trigger internal physiology adjustments that are needed to optimally respond to those changing underwater pressures.

> It turns out that the seals follow fish trails by sensing very subtle water pressure changes with their whiskers....To do this, seals detect and interpret "the structure and spatial arrangement of the vortices" that spin off from a fish's underwater trail. Not only can seals detect the vortices, but they can sense the "high water velocities" behind a swimming fish even after the fish is long gone. Water that trails a fish flows just a little faster than the surrounding waters.[6]

As warm-blooded mammals, harbor seals need to burn food energy to keep warm. Blubber helps insulate the seal's core, but food energy is a must, constantly. So, to maintain their body temperature, especially while they swim in super frigid seawaters, they must eat a lot. They are habitually hungry!

Harbor seals don't really "chew" their food, though. Rather, they bite and tear, crush with their molars, then quickly swallow. Harbor seals frequently feast on cod, sea bass, mackerel, anchovy, whiting, herring, hake, sole, flounder, some crustaceans (including crabs and shrimps), small-sized octopus or squid, sometimes salmon or trout, or maybe even a sea-duck. They eat about 5% of their body weight each day!

Of course, seals have a right to be hungry, and to eat a lot, because God made them to operate that way. They have a lot of work to do as they eat and swim and dive, always displaying God's marvelous bioengineering.

References

1. "O LORD, how manifold are Your works! In wisdom You have made them all. The earth is full of Your possessions—this great and wide sea, in which are innumerable teeming things, living things both small and great." (Psalm 104:24–25).

2. Whitaker, J. O., Jr. 1998. *National Audubon Society Field Guide to North American Mammals*, revised edition. New York, NY: Alfred A. Knopf, 728–730, Plate # 358.

3. Dr. Gro Villanger and her team reported: "PFOS [perfluorooctane sulfonate] concentrations in white whales were about half the concentrations in harbour (*Phoca vitulina*) and ringed (*Pusa hispida*) seals, similar to hooded seals (Cystophora cristata) and higher than in walruses (*Odobenus rosmarus*) from that same area. From 1996…to 2013-2014, plasma concentrations of PFOS decreased by 44%, whereas four C9-12 PFCAs [perfluoroalkyl substances] and total PFCAs increased by 35-141%. These results follow a similar trend to what has been reported in other studies of Arctic marine mammals from Svalbard." See Villanger, G. D. et al. 2020. Perfluoroalkyl Substances (PFASs) in White Whales (*Delphinapterus leucas*) from

Svalbard: A Comparison of Concentrations in Plasma Sampled 15 Years Apart. *Environmental Pollution.* 263 (A): 114497.
4. Magnusson M. and H. Pálsson, trans. 1982. *Njal's Saga,* chapter 19. New York, NY: Penguin Classics.
5. Vancouver Aquarium Marine Science Center. 2005. AquaFacts: Harbour Seals (*Phoca vitulina*), pages 1–4. Citin Kleene, S. 1989. A Medical Marvel: the Diving Seal. *Sea Frontiers.* 35: 370–374.
6. Thomas, B. Seal Whiskers Track Fish Trails. *Creation Science Update.* Posted on ICR.org June 22, 2010, accessed May 1, 2020.

21
COMPLEX METABOLIC PROCESS IN FISH STARTLES EVOLUTIONISTS

Jeffrey P. Tomkins, Ph.D.

Autophagy is an amazingly complex and ingenious process in which cells are able to degrade and recycle their own damaged or dysfunctional components. It not only allows for the efficient recycling of important molecules and biochemical structures but also produces a stable equilibrium between interdependent cellular elements and physiological processes. In other words, it's essential to life.

There are actually a variety of biochemical pathways for autophagy in the cell. One of these is called chaperone-mediated-autophagy (CMA). It contributes to the degradation and recycling of a wide range of proteins and is essential in regulating various cell functions, including gene activity, DNA repair, cell cycle regulation, and cellular metabolism. Mutations in genes regulating CMA have been linked to a number of human diseases, including neurodegenerative disorders, various types of cancer, metabolic problems, and immune system dysfunction.

Up until recently, scientists have believed that

Medaka

CMA was a recent evolutionary development that first appeared in land vertebrates, namely mammals and birds. This was due to researchers' prior inability to locate genes associated with CMA in fish. And quite conveniently, this scenario fit well with the evolutionary claim that fish are the primitive ancestors of land vertebrates and would thus supposedly lack more advanced cellular systems.

More specifically, the evolutionists were basing their assumptions on the absence of any identifiable LAMP2A gene in fish. This gene encodes a necessary protein for CMA activity in mammals and birds. However, in 2018, researchers finally stumbled across some evidence for the gene in a fish database called Phylofish.[2] They discovered that several fish species had active genes that were fairly similar to the mammalian version of LAMP2A. This finding startled them because it suggested that CMA was active in fish much earlier in vertebrate evolution than initially thought. This serves as another example that demonstrates how a lack of human knowledge is not evidence for evolution.

In a 2020 study, these same researchers bolstered the evidence for LAMP2A and CMA activity in fish.[1] By using specialized fluorescent biomolecule reporting technology that can be visualized with a light microscope, they confirmed the existence of a CMA pathway in cells of the fish medaka (*Oryzias latipes*). They also mutated the LAMP2A gene in fish, which resulted in severe dysfunction in carbohydrate and fat metabolism—a result previously demonstrated in mice. Taken together, this new study shows that CMA also occurs in fish where it performs an essential role in cell metabolic regulation.

This research also shows that the diverse complexity of autophagy systems was present in vertebrate systems fully formed and functional with no evolutionary precursors. While these results confound evolutionary ideas and presuppositions about life steadily progressing from simple to complex, the data fit perfectly with the creation model of plant and animal origins where complex cellular systems have been present from the very beginning and can be found at all levels of life.

References

1. Lescat, L. et al. 2020. Chaperone-Mediated Autophagy in the light of evolution: insight from fish. *Molecular Biology and Evolution*. 37 (10): 2887–2899.
2. Lescat, L. et al. 2018. CMA restricted to mammals and birds: myth or reality? *Autophagy*. 14 (7): 1267–1270.

22
FISH BODY DESIGN REVEALS OPTIMIZED SWIMMING MECHANICS

Jeffrey P. Tomkins, Ph.D.

Engineering-minded scientists have taken notice that many types of fish have bodies shaped like a low-drag airfoil that is characteristic of airplane wings. Now, a 2020 research study has proven that the engineered mechanics of this design in fish provide optimized movement and thrust for swimming.[1]

Human-designed airplane wings have a rounded leading edge with a smoothly tapered trailing section that is uniquely shaped to reduce drag while moving through the air at high speeds. Over the years, scientists have been studying the design of both birds and fishes to maximize human-engineered systems for moving airplanes rapidly and efficiently.[2] However, researchers wanted to know more about how fish body design enabled them to use their unique shape to efficiently produce high levels of thrust during high-speed propulsion. They were already aware that fish were able to use this design for highly efficient, leisurely cruising.

Compared to the wing of an airplane, the head-

Bluegill sunfish

to-tail body plan of many fishes is shaped similarly to an airfoil turned on its side. When a fish swims, it oscillates in such a way that it produces negative pressure along the length of its body due to its shape and pitching movements. The negative pressure produces maximal thrust forces along the entire length of its body, but scientists did not understand the exact nature of this design. In this study, researchers applied a high-resolution, pressure-based analysis to two types of fishes: a bluegill sunfish and a brook trout.

The fish were shown to produce thrust toward the front of their bodies using leading-edge suction mechanics similar to an airfoil. In fact, the swimming motion that produced thrust on the frontal portions of the fish bodies through negative pressure yielded 28% of the total thrust produced over the whole body. And incredibly, this specific percentage was also optimal for decreasing the net drag on the trailing part of the body. The end result was an amazingly optimized design based on the known principles of fluid mechanics. In their adulation of the optimally efficient design, the authors stated,

> We suggest that such airfoil-like mechanics are highly efficient, because they require very little movement and therefore relatively little active muscular energy, and may be used by a wide range of fishes since many species have appropriately shaped bodies.[1]

Unfortunately, their analysis did not consider the obvious implications of what the Bible says in Psalm 19:1–2, "The heavens declare the glory of God; and the firmament shows His handiwork. Day unto day utters speech, and night unto night reveals knowledge." This highly optimized handiwork of the Lord leaves those who deny the Creator, the Lord Jesus Christ, without excuse.

References

1. Lucas, K. N., G. V. Lauder, and E. D. Tytell. 2020. Airfoil-like mechanics generate thrust on the anterior body of swimming fishes. *Proceedings of the National Academy of Sciences.* 117 (19): 10585–10592.
2. Floryan, D., T. Van Buren, and A. J. Smits. 2018. Efficient cruising for swimming and flying animals is dictated by fluid drag. *Proceedings of the National Academy of Sciences.* 115 (32): 8116–8118.

23
DOES ODDBALL PLATYPUS GENOME REVEAL ITS ORIGINS?

Brian Thomas, Ph.D.

How in the world did a creature as odd as the duck-billed platypus originate? This creature lays eggs like a reptile, has venom like a reptile, spurs like a chicken, excretes milk from belly patches to nurse its young, has fur that glows, webbed feet like a duck, and uses its sensor-filled, duck-like bill to find aquatic prey like paddlefish do since it swims with its eyes closed. If it evolved, then did its ancestors include mammals, reptiles, birds, and fish? Researchers looked to its genetics to help unravel its origins. In the end, their conclusions drew more from philosophy than fact.

The research team based at the University of Copenhagen published their findings in the prestigious journal *Nature*.[1] They used multiple sequencing techniques to capture different lengths of platypus DNA. Computational methods stitched these lengths together and assigned them to positions on the platypus chromosomes. They compared various DNA sequences to similar ones in chicken,

Platypus

Tasmanian devil, common wall lizard, opossum, and human DNA.

The group did plenty of work gathering interesting data. But what do the DNA sequences mean? Professor Guojie Zhang of the Department of Biology at Copenhagen said, "The complete genome has provided us with the answers to how a few of the platypus' bizarre features emerged."[2]

Answers like what? Zhang said, "Genetically, it is a mixture of mammals, birds and reptiles." But genes code for traits, and we already saw the platypus traits that resemble those groups. So, this statement offers no new answers.

The team zoomed in on the uniqueness of egg-laying in mammals. Today, chicks get all their nutrients from within the egg before they hatch. Newly-hatched puggles get some nutrients from within their platypus eggs, but they still need to lap mother's milk. And of course, human babies initially get all their nutrients from milk. The research group demonstrated that chickens have three egg-producing genes, the platypus has one, and humans have none.

They concluded that since all three evolved from a common ancestor that laid eggs, the platypus lost two egg-specific genes while humans lost all three. This conclusion relies entirely on the premise of a common ancestor. It simply ignores the at least equally logical divine origins option. A smart Creator could have equipped each of these three creatures with the specific DNA sequences needed to carry out its own unique growth and development.

The team also focused on genes that handle the platypus' unique blood system. In most mammals and humans, three specific molecules routinely manage heme and hemoglobin recycling. Heme molecules left to drift within body systems would kill. So, animal life depends upon proper heme processing. But the platypus is missing one of those three management molecules that other mammals typically have. The study authors suggest that it uses a different molecule.

Chickens use a CD163-type molecule in their blood processing, and the platypus has its own version of that—presumably doing something—in its own body cells. The *Nature* study authors wrote that "CD163 family protein(s) may have evolved this role [heme management] in monotremes."[1] But how could this unique platypus blood-processing system have evolved if the animal couldn't survive without this function to begin with?

The statement that platypus heme management proteins evolved commits a circular reasoning fallacy. The statement ignores the irreducible complexity of blood processing. What molecules processed blood while the CD163 proteins were supposedly evolving "this role in monotremes"? The creatures would have immediately died without its blood and blood managers fully functional from the start.

Platypus gene sequencing contributes knowledge to science but does not help explain how these creatures supposedly evolved. Conclusions that presume their own premises fail to reach the status of data-driven science. On the other hand, the fully formed genome that the platypus does possess shows the exact kind of life-giving organization that one would expect from intentional craftsmanship.

References

1. Lucas, K. N., G. V. Lauder, and E. D. Tytell. 2020. Airfoil-like mechanics generate thrust on the anterior body of swimming fishes. *Proceedings of the National Academy of Sciences.* 117 (19): 10585–10592.
2. Floryan, D., T. Van Buren, and A. J. Smits. 2018. Efficient cruising for swimming and flying animals is dictated by fluid drag. *Proceedings of the National Academy of Sciences.* 115 (32): 8116–8118.

24
BEETLE MOUTH GEARS SHOUT DESIGN

Frank Sherwin, D.Sc. (Hon.)

Beetles (order Coleoptera) are a unique but common group of insects easily recognized by the pair of shiny forewings covering their body. These protective wing-cases are called the elytra. Beetles make up almost 40% of the described insects in God's creation.[1] If all zoologists stopped what they were doing and investigated just creatures in Coleoptera, they would easily be busy well into the next century.

Beetle research continues to amaze. Japanese biologists recently discovered an astonishing structure within the mandibles (mouth pincers) of the horned beetle (subfamily Dynastinae, also known as the Rhinoceros beetle). The insects were discovered to have complex, gear-like structures that operate in "completely synchronous movements."[2] Entomologists at the Tokyo University of Agriculture and Technology reported, "A closer examination revealed that each mandible has two gear teeth, and the two sets mesh. As a result, when one mandible moves, so does the other."[3]

In 2013, *New Scientist* magazine reported unexpected machinery in insect larvae:

> The insect *Issus coleoptratus* is another animal with an unexpected bit of machinery hidden in its body. Its larvae are the first animals known to have interlocking gears, just like in the gearbox of a car.[4]

Did these complex movements and gear teeth come about by time and chance or by plan and purpose? Evolutionists would immediately explain that the gears were not designed but somehow slowly developed piecemeal. However, the 2013 *New Scientist* article stated, "It might be that gears are easily broken, and as soon as one tooth is sheared off, the mechanism doesn't work as well."[4] If that's true, then how could such a mechanism evolve from teeth that just randomly appeared, one after the other? A partial gearbox in a car certainly wouldn't work, nor would living gears that just randomly intermesh and rotate.

Rhinoceros beetle

The synchronous movements of the horned beetles' mouth gears shout design, plan, and purpose!

References

1. Stork, N. E. 2015. New approaches narrow global species estimates for beetles, insects, and terrestrial arthropods. *Proceedings of the National Academy of Sciences.* 112 (24): 7519–7523.
2. Ichiishi, W. et al. 2019. Completely engaged three-dimensional mandibular gear-like structures in the adult horned beetles reconsideration of bark-carving behaviors (Coleoptera, Scarabaeidae, Dynastinae). *ZooKeys.* 813: 89–110.
3. Marshall, M. 2019. Rhinoceros beetles have weird mouth gears that help them chew. *New Scientist.* Posted on newscientist.com February 1, 2019.
4. Marshall, M. 2013. Zoologger: Transformer insect has gears in its legs. *New Scientist.* Posted on newscientist.com on September 12, 2013, accessed on February 12, 2019.

Termite mound in northern Australia

25
TERMITE NEST ARCHITECTURAL DESIGN IS CLEARLY SEEN

Frank Sherwin, D.Sc. (Hon.)

Termites (order Isoptera) are eusocial—animals with an advanced social organization—insects that can number in the millions, producing something biologists call a superorganism. A superorganism is a colony of termites having features of organization analogous to the properties of a single creature.

> Creationists maintain that termites have always been termites. Looking to the fossil record shows them in sediments dated by evolutionists to be "251 million years old"[1]— but they're still termites as they exist today. Other evolutionists think they originated in the Carboniferous about "359 million years ago." Regardless, no ancient common termite ancestor has been discovered.

Some termites set up a small nest in the foundation of an American home, since these insects can digest wood due to symbiotic gut-dwelling flagellates, single-celled eukaryotic creatures that have enzymes capable of breaking down cellulose. Others, though,

Termites

build formidable mounds (termitaria) in Australia or West Africa. They use their saliva, dung, and soil to construct amazing mounds that are temperature and moisture-controlled cities for the colony. Words such as "engineering," "mechanisms," and "design" are constantly used in articles describing termite nest construction.

For example, with millions of termites living in a small area there is bound to be an increase in CO_2 levels, with additional CO_2 contribution from fungus cultivation.[2] The accumulation of this gas can be toxic unless it's dissipated to the outside through some kind of ventilation. In addition, there must be thermal insulation control. But how are such controls achieved? Good engineering design, of course.

Atmosphere and CO_2 levels are exchanged via many thousands of millimeter-sized, external "windows" in the mound's outer wall. The termites have been designed with the ability to frequently open and close these tiny windows based upon outside breezes and CO_2 accumulation in the nest. But it wouldn't be any good—or it may even be counterproductive—for termites to randomly open and close the windows. The numerous termites must all operate as a unit—a superorganism. Did such environmentally-induced behavior come about by time and chance or plan and purpose?

In the article, the authors ask the following regarding one of two kinds of termite nests.

> If the outer wall is porous, then are the pores connected and permeable? And if so, how do they contribute to air circulation or ventilation? In addition, does the porous structure of walls control the thermal insulation and the structural stability of the nest?[2]

These are overtly engineer-driven questions, which logic dictates will result in engineer-based answers.

> The network of larger microscale pores enhances permeability by one to two orders of magnitude compared to the smaller pores alone, and it increases CO_2 diffusivity up to eight times. In addition, the pore network offers enhanced thermal insulation and allows quick drainage of rainwater, thereby restoring the ventilation and providing structural stability to the wet nest.[2]

Once the construction and thermoregulation questions are answered, then the knowledge gained can be applied to human building design.

The mounds provide a self-regulating living environment that responds to changing internal and external conditions. A multidisciplinary team of engineers and entomologists is looking at whether similar principles could be used to design buildings that need few or no mechanical services (e.g., heating and ventilation) and so use less energy and other resources than conventional structures.[3]

Evolution's unreliable components of chance and time simply did not contribute to the design seen in these sophisticated and highly successful nests.

References

1. Weesner, F. M. 1960. Evolution and Biology of the Termites. *Annual Review of Entomology.* 5 (1): 153–170.
2. Singh, K. et al. 2019. The architectural design of smart ventilation and drainage systems in termite nests. *Science Advances.* 5 (3): eaat8520.
3. Termites could hold the key to self-sufficient buildings. *EurekAlert!* Posted on eurekalert.org September 21, 2004, accessed March 27, 2019.

26
IMPROVED STEEL COPIES BONE MICROSTRUCTURE

Jeffrey P. Tomkins, Ph.D.

How does one build a structural material that withstands stress and fracture? The answer is to copy optimal designs from living systems because they far exceed man's ingenuity. Recently, an improved steel was developed by copying human bones.[1]

Human and animal bones are optimally designed to be lightweight, incredibly strong, and resistant to fracture and fatigue. These bones are also self-healing and fully integrated—both physiologically and structurally—with the rest of the body. Bones are excellent examples of God's creative genius.

The amazing bone features are facilitated by an incredibly ingenious microstructure that's engineered at hierarchical levels.[2] When observed at the nanoscale level, a network of tiny fibers composed of collagen, a

Bone cutaway

127

type of protein, are found in an intricately layered arrangement. Fibers are oriented in different directions in each layer. On a slightly larger scale, bones exhibit a framework with a lattice-like structure that makes them both strong and light. This type of multi-level engineering ensures that bones resist cracking in any one particular direction—a feature that materials engineers have long sought.

Bone structure detail

The mechanics of crack propagation is an important area of research for things like cars and skyscrapers. The properties of such materials can, in some cases, determine human life and safety. Because of this, materials that resist cracking have been studied by scientists (called metallurgists) for many years.

The main problem for metallurgists is the attainment of both strength and toughness in a single engineered material, because these properties are typically mutually exclusive. Materials that are really strong (hard) tend to be highly subject to fracturing—they tend to be brittle. Materials that are tough (fracture resistant) tend to be pliable and flexible. When developing materials that are both strong and tough, there is often an uneasy compromise somewhere. In their quest to achieve this seemingly unattainable engineering feat, researchers copied the Creator's divine engineering they found in bones and published their findings in the prestigious journal *Science*.[1]

The Bible says, "For since the creation of the world His invisible attributes are clearly seen, being understood by the things that are made" (Romans 1:20). This news story perfectly highlights this profound scriptural truth. Here we have an age-old engineering problem that humans have been unable to solve, but with the modern tools of microstructure analysis, metallurgists are able to behold the incredible structural engineering performed by our great Creator God and then copy it, albeit at a crude level.

Clearly, discoveries like this do not glorify mankind but rather our omnipotent Creator. He deserves all the credit.

References

1. Koyama, M. et al. 2017. Bone-like crack resistance in hierarchical metastable nanolaminate steels. *Science*. 355 (6329): 1055–1057.
2. Tertuliano, O. A. and J. R. Greer. 2016. The nanocomposite nature of bone drives its strength and damage resistance. *Nature Materials*. 15 (11): 1195–1202.

27
DESIGN-BASED SPIDER RESEARCH PROVES CREATOR'S GENIUS

Jeffrey P. Tomkins, Ph.D.

Evolutionary theory is based on the faulty assumption that random processes can produce highly ordered, complex systems, and this theory routinely fails to satisfactorily and sufficiently explain scientific reality. For example, a 2019 conventional study that applies engineering principles in the analysis of spider-web physics, related to the design concept of "power amplification," has been highly successful, making evolutionary explanations unbelievable. Instead, it glorifies the amazing genius of God, the Creator.[1]

Power amplification is an ingenious engineering principle that enables animals to produce exceptionally rapid and powerful movements that exceed the normal physiological limits of muscle power and speed. Some of the best known examples of this design principle in nature are the ultrafast strike of the mantis shrimp's claw, a power-packed thrust that can break aquarium glass, and the flea's jump, in which it accelerates at 100 Gs (force of gravity) while

Triangle-weaver spider's web. Hey, it's shaped like a triangle…imagine that!

reaching heights of over 100 times its body length.[2,3] However, until now, scientists have only been aware of examples of muscle-driven power amplification that involve anatomical structures alone, like the above examples that produce rapid bursts of movement from a single cycle of muscular contraction.

This 2019 study focused on the triangle-weaver spider (*Hyptiotes*) that cleverly uses its web as a crafty tool to capitalize on the principle of power amplification. In an amazing confluence of innate creature behavior and engineering design, the spider produces an ultra-rapid, prey-entangling web action unlike a typical spider's web that is sticky but static. The researchers in this study commented that this is "the only known example of external power amplification outside of human tools."[1]

Triangle-weaver spider

The spider expertly generates, and eventually releases, tension by waiting for prey in the corner of its triangular-shaped web, where main radial threads converge with its body. Its body acts as a link between an anchor line connected to cross threads and a trap line connected to the main triangle of the web. In preparing for action, the spider moves backward along the anchor line in a unique leg-over-leg motion that pulls the springy web taut, loading it with immense energy over multiple tightening cycles.

When a prey lands on the web, the spider releases its grip on the anchor line, and both the spider and web instantly surge forward. The action also causes multiple lines of spider silk to slam into and fully entangle the prey. Any further forward movement ceases as slack is removed from the anchor line by the spider squeezing its internal spinnerets so that no more silk is produced. If the prey is exceptionally large or vigorous, the spider can repeat its unique actions multiple times to tighten and release the lines to further entangle prey.

The study authors discovered that the elastic energy stored in the web and expertly manipulated by the spider creates a huge amount of propulsive force, as opposed to simply jumping or moving the web with its legs like other spiders. As a result of these new observations, we have another unique example of a creature generating enormous power that exceeds muscular capacity alone.

The researchers who analyzed the amazing physics and energy kinetics behind this spidery genius claimed an analogy to human-engineering. They stated,

> This finding reveals an underappreciated function of spider silk and expands our understanding of how power amplification is used in natural systems, showing remarkable convergence with human-made power-amplifying tools.

Such complex, design-based ingenuity in biological systems, especially one where both highly intelligent creature behavior and complicated bioengineering converge, ultimately glorify the mighty Creator who made it all.

References

1. Han, S. I. et al. 2019. External power amplification drives prey capture in a spider web. *Proceedings of the National Academy of Sciences.* 116 (24): 12060–12065.
2. Patek, S. N., W. L. Korff, R. L. Caldwell. 2004. Biomechanics: Deadly strike mechanism of a mantis shrimp. *Nature.* 428 (6985): 819–20.
3. Sutton, G. P., M. Burrows. 2011. Biomechanics of jumping in the flea. *Journal of Experimental Biology.* 214 (5): 836–847.

Spider webs display precise geometric design

28
SPIDERS HAVE BUILT-IN ALGORITHM TO CONSTRUCT WEBS

Jeffrey P. Tomkins, Ph.D.

One of the many mysteries of biology is how a creature like a web-weaving spider with a tiny brain can systematically construct an elaborate web with amazing elegance, complexity, and exacting geometric precision. And to make the spidery task even more amazing, the creature does it blindly, only using the sense of touch—an exceedingly complex mechano-sensory-based application. A 2021 study shows that this remarkable skill is due to a highly sophisticated, built-in algorithm.[1]

To understand how the tiny brains of spider architects enable their sophisticated web construction, the first logical step is to systematically document and analyze their web-weaving behaviors and the specific motor skills of all their moving parts. However, this is no easy task since spiders often do this in the dark with the coordinated specificity of eight rapidly moving legs.

In this study, researchers analyzed a spider native to the western United States called the hackled orb weaver. This spider is small enough to sit comfortably on the tip of a human finger. Because the hackled orb weaver does its work at night, the researchers designed an experiment using infrared cameras and infrared lights. Then, using high-speed cameras, they monitored and recorded six spiders every night as they constructed their webs. The video data allowed the scientists to track literally millions of individual leg movements with sophisticated machine vision software. The researchers had to train the software to detect the body and leg posture of the spiders, frame-by-frame, to document the entire repertoire that the spiders' legs performed to build a complete web.

One of the key discoveries from the huge amount of data was that the web-making behaviors were similar across the six separate spiders. In fact, the data were so homogenous that the researchers could determine the specific part of a web that a spider was constructing just from observing the position of its legs. In other words, even if the final overall web structure was slightly different, the rule-based algorithm each spider used to build a web were the same. In an interview, one of the scientists commented that "They're all using the same rules, which confirms the rules are encoded in their brains."[2]

It is well documented in scientific literature that creatures with tiny brains contain exceedingly complex algorithms that boggle the human mind.[3-6] But the obvious question is: Where did this extreme source of mathematical information originate? The clear answer is that an omnipotent, all-knowing Creator, the Lord Jesus Christ, engineered these systems into the diversity of creatures He made.

References

1. Corver, A. et al. 2021. Distinct movement patterns generate stages of spider web building. *Current Biology.* 31 (22): 4983–4997.
2. Rosen, J. Spiders' web secrets unraveled. *ScienceDaily.* Posted on sciencedaily.com November 1, 2021, accessed November 16, 2021.
3. Tomkins, J. P. Communal Nutrition in Ants: Strong Evidence for Creation. *Creation Science Update.* Posted on ICR.org July 8, 2009, accessed November 16, 2021.
4. Thomas, B. 2010. Bees Solve Math Problems Faster Than Computers. *Creation Science Update.* Posted on ICR.org November 2, 2010, accessed November 16, 2021.
5. Thomas, B. Scientists Discover the 'Anternet'. *Creation Science Update.* Posted on ICR.org September 14, 2012, accessed November 16, 2021.
6. Tomkins, J. P. Ant Behavior Informs Computer Search Algorithms. *Creation Science Update.* Posted on ICR.org June 22, 2020, accessed November 16, 2021.

29
DO SHRINKING SHREWS CHEAT EVOLUTION?

Jeffrey P. Tomkins, Ph.D.

Common shrews are uniquely engineered creatures that have a high metabolism—very different from your average mammal. And now biologists have just discovered the shrew's built-in adaptive secret to over-wintering that utterly defies the standard evolutionary paradigms.[1]

Common shrews exhibit one of the highest levels of body metabolism among mammals. As a result of their high energy requirements, they are voracious little eaters and consume a large amount earthworms and insects for their small size. Because whatever fat reserves they might have are quickly used up, they can starve to death after being deprived of food for only a few hours. Despite this unusual biology and lifestyle, shrews are extremely successful critters and are very widespread across the Northern Hemisphere. Needless to say, this whole scenario baffles the minds of evolutionists.

But the shrew becomes even more fascinating when we consider that, unlike many other animals, shrews neither store up food nor hibernate during winter. Instead, they grow rapidly during the spring and summer months, but in late fall, their bodies begin to shrink. They eventually lose up to 30% of their total body weight. In addition to reduction of fat and muscle mass, internal organs also shrink, including their brains and skulls.[2] Then in the early spring, the shrews start to grow again.

To the evolutionary biologist, this whole strategy appears to be a paradox. Despite having a winter coat of fur, one would expect the little shrews to cool off quickly in the low winter temperatures. This is because small animals like shrews have a seemingly unfavorable body surface to body mass ratio and tend to lose more heat. Outside the strange world of shrews, the relationship between a creature's body weight, metabolic rate, and air temperature is considered a fundamental paradigm of ecology.

In a 2020 study, the scientists set out to determine how the seasonal change in body size affected the

Common shrew

shrews' energy consumption.[1] They did this by measuring metabolism in outdoor temperatures across the various seasons. The results were startling. The senior author of the study claimed, "The common shrew somehow manages to cheat evolution."[3] So, are the shrews really evolution cheaters or amazingly designed adaptive marvels?

This study showed that they actually do not consume more energy per unit of body weight during the winter months despite environmental temperature fluctuations of more than 30° and their small body size. Amazingly, this unique adaptation is not because they are less active during winter. When the researchers analyzed video recordings, the shrews appeared to rest only slightly more, but this did not explain the major differences in energy requirements.

When all the data were combined, the scientists determined that the shrews consistently produced high levels of heat because of their regular high metabolic rates. Therefore, they did not need to increase their metabolic rates during winter because complex adaptive reductions in body size enabled them to consume less overall energy. This is certainly helpful considering the scarcer shrew food supply (fewer insects) in winter.

Once again, amazing and creative adaptive design is the only obvious conclusion, and this points directly to the omnipotent Creator who engineered all life on Earth.

References

1. Schaeffer, P. J. 2020. Metabolic rate in common shrews is unaffected by

temperature, leading to lower energetic costs through seasonal size reduction. *Royal Society Open Science.* 7 (4): 191989.
2. Lázaro, J. et al. 2017. Profound reversible seasonal changes of individual skull size in a mammal. *Current Biology.* 27 (20): R1106–R1107.
3. Max Planck Society. Shrinking instead of growing: How shrews survive the winter. *Phys.org.* Posted on phys.org April 28, 2020, accessed April 28, 2020.

30
HERO SHREW SPINE DESIGN GLORIFIES THE CREATOR

Jeffrey P. Tomkins, Ph.D.

When you first look at a *hero* shrew, you might wonder, "How in the world did this critter get this name?" But these little, mole-like creatures are considered the Clark Kents of the animal world—their superpowers are hidden under humble exteriors. New research into the amazing structure and function of the hero shrew spine is revealing amazing engineering that utterly defies evolutionary speculation.[1]

Nothing else in the animal kingdom contains the extreme spinal structure and strength of hero shrews.[2] Their highly unique and unusual backbones can withstand the weight of a full-grown human standing on top of them. The hero shrew's spine has interlocking vertebrae that make it extremely strong and rigid when it is compressed. Interestingly, the shrew's backbone is flat on both the top and underneath. It also features plentiful amounts of broad side flanges that have lots of finger-like projections forming a nearly circular cage around the entire spine.

Hero shrew

To more fully ascertain how the design of this structure gives the hero shrew its amazing spinal powers, scientists used a 3-D X-ray technology to scan the internal features of vertebrae using 20 different specimens.[1] For a comparison, they also included scans from a goliath shrew that is similar in size to the hero shrew, but the goliath shrew has a standard mammalian backbone. The goal was to analyze the density and cellular structure of the interior of the bones and then combine that analysis with the morphology data of the outward mechanics of the spine.

Compared to the spine of goliath shrews, hero shrews have many more and much-wider vertebrae. Internally, the hero shrew's cellular structure exhibited many reinforcing, rod-like structures that made the otherwise spongey bone very dense and strong. These struts were primarily oriented in a head-to-tail direction. This is different from the goliath shrew where the struts seemed to have a more random orientation. This strut orientation within the bone tissues of a hero shrew further maximize the strength and power of the spine.

The bottom line is that hero shrew spines don't just look tough and formidable from the outside. Their inside cellular structure is also arranged to maximize strength. Optimization is a hallmark of design.

What purpose this unique design is serving the shrew is yet to be determined since little is known about the behavior of these creatures in the wild. One hypothesis is that this trait facilitates foraging for food by allowing the shrew to scrunch up and force its way through difficult places to get at insects for lunch.

Unique adaptive designs in shrews and other creatures continue to amaze biologists. And, of course, evolutionists have no clue as to how these highly complex and optimized all-or-nothing traits could have appeared suddenly with no evolutionary precursor. The obvious conclusion is that an all-powerful Creator engineered these creatures from the beginning.

References

1. Smith, S. M. and K. D. Angielczyk. 2020. Deciphering an extreme morphology: bone microarchitecture of the Hero Shrew backbone (Soricidae: Scutisorex). *Proceedings of the Royal Society* B. 287: 20200457.
2. Stanley, W. T. et al. 2013. A new hero emerges: another exceptional mammalian spine and its potential adaptive significance. *Biology Letters*. 9: 20130486.

31
GECKOS HAVE HOLES IN THEIR HEADS
Frank Sherwin, D.Sc. (Hon.)

The lovable gecko made the news again in 2018.[1] In 2009, it was discovered the gecko has amazing nocturnal vision.[2] Even their sophisticated feet that secrete phospholipids, complex membrane molecules, are a wonder of creation.[3]

The impediment of directional hearing in small animals, such as the gecko, is cleverly solved by the Creator's design. In larger creatures, the location of noises is resolved by a procedure called triangulation. It's a method of determining something's location by measuring angles from known points and using the location of other things via a fixed baseline. In humans, our pinna (outer ear) of each ear is designed to hear in stereo. This allows our brain to triangulate and thus discern where sounds are coming from.

However,

Geckos and many other animals have heads that are too small to triangulate the location of noises the way we do, with widely spaced ears. Instead, they have a tiny tunnel through their heads that measures the way incoming

Madagascar gecko

sound waves bounce around to figure out which direction they came from.[1]

It has been discovered that these fascinating creatures are designed with one eardrum that

> essentially steals some of the sound wave energy that would otherwise tunnel through to the other. This [interference] helps the gecko—and about 15,000 other animal species with a similar tunnel—understand where a sound is coming from.[1]

The created gecko, and its sophisticated directional hearing on such a small scale, was the subject of a *Nature Nanotechnology* paper entitled, "Subwavelength angle-sensing photodetectors inspired by directional hearing in small animals."[4] Researchers from the University of Wisconsin-Madison and Stanford University lined up two tiny silicon wires called nanowires (about 1/1,000th as wide as a hair) in a manner that mimics the gecko's eardrums. Through this positioning they were able to "map the angle of incoming light waves" in the light-detector experiment they were conducting.[1]

A graduate student involved in the research said that geckos were not the inspiration for the original assembly of this light system. They "came upon the likeness between their design and geckos' ears after the work had already begun." But there was a significant level of similarity: "The same math that explains both the gecko ears and this photodetector describes an interference phenomenon between closely arranged atoms as well."[1]

Clearly God's minute design features in His living creation have substantiated the direction and quality of this light-detector system.

References

1. Stanford University. Tiny light detectors work like gecko ears. *ScienceDaily*. Posted on sciencedaily.com October 30, 2018, accessed November 15, 2018.
2. Thomas, B. Gecko Eyes Make Great Night Vision Cameras. *Creation Science Update*. Posted on ICR.org May 29, 2009, accessed November 15, 2018.
3. Thomas, B. Scientists Discover New Clue to Geckos' Climbing Ability. *Creation Science Update*. Posted on ICR.org October 17, 2011, accessed November 15, 2018.
4. Yi, S. et al. 2018. Subwavelength angle-sensing photodetectors inspired by directional hearing in small animals. *Nature Nanotechnology.* 13: 1143–1147.

CONTRIBUTORS

Randy J. Guliuzza is the president of the Institute for Creation Research. He earned a doctor of medicine degree from the University of Minnesota, a master of public health degree from Harvard University, and received an honorary doctor of divinity degree from Southern California Seminary. He served in the U.S. Air Force as 28th Bomb Wing flight surgeon and chief of aerospace medicine. Dr. Guliuzza is also a registered professional engineer and holds a B.A. in theology from Moody Bible Institute.

James J. S. Johnson is an associate professor of apologetics and the chief academic officer at the Institute for Creation Research.

Frank Sherwin is a science news writer at the Institute for Creation Research. He earned his M.A. in zoology from the University of Northern Colorado and received an honorary doctor of science degree from Pensacola Christian College.

Dr. Brian Thomas is a research scientist at the Institute for Creation Research and earned his Ph.D. in paleobiochemistry from the University of Liverpool.

Dr. Jeffrey P. Tomkins is a research scientist at the Institute for Creation Research and earned his Ph.D. in genetics from Clemson University.

IMAGE CREDITS

t = top, b = bottom

BigstockPhoto: 10, 11, 13, 19–21, 25, 26, 28, 31, 36, 37, 43, 45, 46, 49, 54, 62, 72, 78, 90, 98, 100, 102, 113, 116, 120, 122, 124, 127, 128, 134, 139, 146

Charles J. Sharp via Wikimedia Commons: 51

Dario Sanches via Wikimedia Commons: 66

Fry72, Karel Frydrýšek via Wikimedia Commons: 12t

Hans Hillewaert via Wikimedia Commons: 83

iStockPhoto: cover, 1, 3, 24

Ivo Antušek via Wikimedia Commons: 39

Jet Lowe via Wikimedia Commons: 12b

Joshua Sera via Wikimedia Commons: 87

Judy Gallagher via Wikimedia Commons: 132

Kotaro Negawa via Wikimedia Commons: 110

Marcial4 via Wikimedia Commons: 63

NASA: 33

NOAA Okeanos Explorer Program via Wikimedia Commons: 93

NOAA Photo Library via Wikimedia Commons: 96

Peter Znamenskiy via Wikimedia Commons: 40

Public domain: 131, 143

Robert McMorran, United States Fish and Wildlife Service: 75

Uoaei1 via Wikimedia Commons: 57

THIS BOOK WAS ADAPTED FROM THE FOLLOWING MATERIALS

Sherwin, F. 2017. Architecture and Engineering in Created Creatures. *Acts & Facts*. 46 (10): 10–12.

Tomkins, J. P. 2019 Complex Creature Engineering Requires a Creator. *Acts & Facts*. 48 (8): 14.

Tomkins, J. P. 2019. Intricate Animal Designs Demand a Creator. *Acts & Facts*. 48 (7): 14.

Guliuzza, R. J. Embryonic "Clocks" Mimic Human Construction Schedules. *Creation Science Update*. Posted on ICR.org February 13, 2020.

Sherwin, F. Honeybee Design Saves Energy. *Creation Science Update*. Posted on ICR.org June 24, 2021.

Sherwin, F. The Passive Stealth Wing of the Moth. *Creation Science Update*. Posted on ICR.org August 22, 2022.

Sherwin, F. Fruit Fly Jitters. *Creation Science Update.* Posted on ICR.org November 17, 2022.

Sherwin, F. Aerial Engineering and Physics of the Dragonfly. *Creation Science Update*. Posted on ICR.org May 26, 2022.

Tomkins, J. P. Open Ocean Dragonfly Migration Boggles the Mind. *Creation Science Update*. Posted on ICR.org November 4, 2021.

Thomas, B. 2020. Why Don't Raindrops Bomb Butterfly Wings? *Acts & Facts*. 49 (8): 20.

Tomkins, J. P. Butterfly Wing Design Repudiates Evolution. *Creation Science Update*. Posted on ICR.org February 18, 2021.

Sherwin, F. 2020. Hummingbirds by Design. *Acts & Facts*. 49 (11): 17–19.

Sherwin, F. Hummingbird Flight Strategies. *Creation Science Update*. Posted on ICR.org January 29, 2024.

Sherwin, F. 2019. The Syrinx Song. *Acts & Facts*. 48 (8): 15–16.

Sherwin, F. 2021. Created Cuttlebone's Sweet Spot. *Acts & Facts*. 50 (2): 15.

Sherwin, F. Molecular Motors of a Squid Show CET in Action. *Creation Science Update*. Posted on ICR.org July 24, 2023.

Sherwin, F. 2021. Marine Sponges Inspire. *Acts & Facts.* 50 (6): 13.

Thomas, B. Clever Construction in Rorqual Whales. *Creation Science Update.* Posted on ICR.org May 14, 2015.

Sherwin, F. Brittle Stars See with Their Skin. *Creation Science Update.* Posted on ICR.org December 26, 2019.

Johnson, J. J. S. Common Seals Display Extraordinary Bioengineering. *Creation Science Update.* Posted on ICR.org May 5, 2020.

Tomkins, J. P. Complex Metabolic Process in Fish Startles Evolutionists. *Creation Science Update.* Posted on ICR.org June 27, 2020.

Tomkins, J. P. Fish Body Design Reveals Optimized Swimming Mechanics. *Creation Science Update.* Posted on ICR.org May 27, 2020.

Thomas, B. Does Oddball Platypus Genome Reveal Its Origins? *Creation Science Update.* Posted on ICR.org January 21, 2021.

Sherwin, F. Beetle Mouth-Gears Shout Design. *Creation Science Update.* Posted on ICR.org March 12, 2019.

Sherwin, F. Termite Nest Architectural Design Is Clearly Seen. *Creation Science Update.* Posted on ICR.org April 4, 2019.

Tomkins, J. P. Improved Steel Copies Bone Microstructure. *Creation Science Update.* Posted on ICR.org April 6, 2017.

Tomkins, J. P. Design-Based Spider Research Proves Creator's Genius. *Creation Science Update.* Posted on ICR.org July 4, 2019.

Tomkins, J. P. Spiders Have Built-In Algorithm to Construct Webs. *Creation Science Update.* Posted on ICR.org November 29, 2021.

Tomkins, J. P. Do Shrinking Shrews Cheat Evolution? *Creation Science Update.* Posted on ICR.org May 11, 2020.

Tomkins, J. P. Hero Shrew Spine Design Glorifies the Creator. *Creation Science Update.* Posted on ICR.org May 7, 2020.

Sherwin, F. Geckos Have Holes in Their Heads. *Creation Science Update.* Posted on ICR.org December 18, 2018.

ABOUT THE INSTITUTE FOR CREATION RESEARCH

At the Institute for Creation Research, we want you to know God's Word can be trusted with everything it speaks about—from how and why we were made, to how the universe was formed, to how we can know Jesus Christ and receive all He has planned for us.

That's why ICR scientists have spent more than 50 years researching scientific evidence that refutes evolutionary philosophy and confirms the Bible's account of a recent and special creation. We regularly receive testimonies from around the world about how ICR's cutting-edge work has impacted thousands of people with Christ's creation truth.

HOW CAN ICR HELP YOU?

You'll find faith-building science articles in *Acts & Facts*, our bimonthly science news magazine, and spiritual insight and encouragement from *Days of Praise*, our quarterly devotional booklet. Sign up for FREE at **ICR.org/subscriptions**.

Our radio programs, podcasts, online videos, and wide range of social media offerings will keep you up-to-date on the latest creation news and announcements. Get connected at **ICR.org**.

We offer creation science books, DVDs, and other resources for every age and stage at **ICR.org/store**.

Learn how you can attend or host a biblical creation event at **ICR.org/events**.

Discover how science confirms the Bible at our Dallas museum, the ICR Discovery Center. Plan your visit at **ICRdiscoverycenter.org**.

P. O. Box 59029
Dallas, TX 75229
800.337.0375
ICR.org

ICR'S MISSION STATEMENT
ICR EXISTS TO SUPPORT THE LOCAL CHURCH THROUGH...

WORSHIP
- Glorify Jesus Christ by emphasizing in all ICR resources the credit He is due as Creator.
- Oppose the deification of nature by exposing Darwinian selectionism as an idolatrous worldview.

EDIFICATION
- Help pastors lead, feed, and defend their flocks by providing scientific responses to secular attacks on the authority and authenticity of God's Word.
- Change Christians' view of biology by constructing an organism-focused theory of biological design that highlights Jesus' work as Creator.

EVANGELISM
- Defend the gospel by showing how natural processes cannot explain the miracles in the Bible.
- Counter objections to the gospel by equipping believers with Scripture-affirming science.

RESOURCES FROM ICR

ICR's Creation Collection provides a biblical understanding of scientific topics. Written by Ph.D. scientists and other experts, each book focuses on a specific area of research that demonstrates both the unreliability of the evolutionary narrative and the infallibility of Scripture. Watch for upcoming books!

Find out more about these books and other resources at **ICR.org/store**